CLIMATE CHANGE AND GLOBAL WARMING

Avinash Tyagi

RAJAT
PUBLICATIONS
New Delhi, India

CLIMATE CHANGE AND GLOBAL WARMING

This edition is published by Rajat Publications
4740/23, Ansari Road Daryaganj,
New Delhi - 110002

ISBN: 978-81-7880-255-8

Edition 2026

Printed and bound in India

RAJAT
PUBLICATIONS

4740/23, Ansari Road, Daryaganj,
New Delhi- 110002 (India)
Tel:- +91-11-23267924
Mob:- +91-9811484973, +91-9811105389
E-mail: rajat_publication@yahoo.com

Preface

The climate of the Earth has never been stable, least of all during the history and evolution of life on Earth. These prehistoric changes in climate were clearly natural in origin and occurred on a planet inhabited by primitive societies with far smaller populations than at present. Indeed, the regularity of the diurnal and seasonal rhythms of our planet has always been overlain by inter-annual, multi-decadal and millennial variations in climate, over whatever timescale climate is defined. Ecosystems and species have moved, often freely, in response to such past changes and have evolved within this climatic history.

The causes of contemporary and future changes in climate, their rate and their potential significance for ecosystems and for the human species, however, are all notably different from anything that has occurred previously in history or pre-history. Evidence of global warming is plentiful and includes data on satellite-measured sea level change, lengthened growing season and increased precipitation intensities resulting in increased flood risk during winter in the northern hemisphere. The causes are now dominated by human perturbation of the atmosphere and the rate of warming already exceeds anything experienced in the last 10,000 years. It is set to become more rapid than anything experienced in recorded human history and, given the ecological imprint made by our current and growing population of six billion and more, the significance of this prospect for the natural world and for human society is qualitatively different from previously experienced changes in climate.

We face certainly continuing, probably accelerating and possibly unprecedented changes in the Earth's climate over the coming years and decades. These changes in such a fundamental resource for society, and in such a powerful influence on ecological, economic and cultural development, will introduce new challenges for the way we live with and influence climate. As evidence is emerging that some physical and biological systems are already reacting to this human-induced change in climate, and as we know that, at least for some regions and for some communities and ecosystems, climate variability already imposes huge costs, doing nothing is unlikely to be the best option.

Societies need to develop and implement appropriate strategies to reduce the risks associated with a changing climate — to ensure that these changing climatic resources are appropriately exploited and that the adverse impacts of changing climatic hazards are minimised. Mitigation measures are required to reduce global greenhouse gas emissions with the intention of eventually stabilising atmospheric concentrations at some level at which an acceptable dynamic equilibrium could be sustained between climate, ecosystems and human society.

This book is a highly informative source on global warming and climate change—issues that are perhaps the greatest threat to this planet. It describes the process of global climate change, its current and future impacts on human health, and how our societies can lessen those adverse impacts, via adaptation strategies and by reducing greenhouse gas emissions. It will be a valuable reference tool for environmental scientists, activists and scholars in the field.

Editor

Contents

Preface v

1. Global Warming and Climate Change 1
2. Nature of Global Climate Change 11
3. Global Warming 27
4. Role of Ocean in Climate Change 40
5. Ice Ages and Global Warming 46
6. Impact of Climate Change on Biodiversity 52
7. Climate Change and Human Health 85
8. Efffects of Climate Change on Coastal Ecosystems 124
9. Impacts of Climate Change on Forests 160
10. Mitigation of Climate Change 185
11. Protecting the Ozone Layer 204
12. Kyoto Protocol to the United Nations Framework Convention on Climate Change 232

Bibliography 270

Index 272

1

Introduction

Weather is the continuously changing condition of the atmosphere, usually considered on a time scale that extends from minutes to weeks. Climate is the average state of the lower atmosphere, and the associated characteristics of the underlying land or water, in a particular region, usually spanning at least several years. Climate variability is the variation around the average climate, including seasonal variations and large-scale regional cycles in atmospheric and ocean circulations such as the El Niño/Southern Oscillation (ENSO) or the North Atlantic Oscillation.

Climate change occurs over decades or longer time-scales. Until now, changes in the global climate have occurred naturally, across centuries or millennia, because of continental drift, various astronomical cycles, variations in solar energy output and volcanic activity. Over the past few decades it has become increasingly apparent that human actions are changing atmospheric composition, thereby causing global climate change. Earth's climate is determined by complex interactions between the Sun, oceans, atmosphere, cryosphere, land surface and biosphere.

The Sun is the principal driving force for weather and climate. The uneven heating of Earth's surface causes great convection flows in both the atmosphere and oceans, and is thus a major cause of winds and ocean currents. Five concentric layers of atmosphere surround this planet. The lowest layer extends from ground level to around 10-12 km altitude on average. The weather that affects Earth's surface develops within the troposphere. The next major layer extends to about 50 km above the surface. The ozone within the stratosphere absorbs most of the sun's higher energy ultraviolet rays. Above the stratosphere are three more layers: mesosphere, thermosphere and exosphere.

Overall, these five layers of the atmosphere approximately halve the amount of incoming solar radiation that reaches Earth's surface. In particular, certain "greenhouse" gases, present at trace concentrations in the troposphere, absorb about 17% of the solar energy passing through it. Of the solar energy that reaches Earth's surface, much is absorbed and reradiated as long-wave radiation. Some of this outgoing infrared radiation is absorbed by greenhouse gases in the lower atmosphere, which causes further warming of Earth's surface. This raises Earth's temperature by 33°C to its present surface average of 15°C. This supplementary warming process is called "the greenhouse effect".

The 1980s and 1990s were unusually warm. Globally speaking, the decades have been the warmest since accurate records began somewhat over a hundred years ago and these unusually warm years are continuing into the twenty-first century. In terms of global average near-surface air temperature, the year 1998 was the warmest in the instrumental record and the nine warmest years in that record have occurred since 1990. The period has also been remarkable for the frequency and intensity of extremes of weather and climate. For example, periods of unusually strong winds have been experienced in western Europe.

During the early hours of the morning of 16 October 1987, over fifteen million trees were blown down in southeast England and the London area. The storm also hit Northern France, Belgium and The Netherlands with ferocious intensity; it turned out to be the worst storm experienced in the area since 1703. Storm-force winds of similar or even greater intensity but covering a greater area of western Europe have struck since—on four occasions in 1990 and three occasions in December 1999. But those storms in Europe were mild by comparison with the much more intense and damaging storms other parts of the world have experienced during these years.

About eighty hurricanes and typhoons—other names for tropical cyclones—occur around the tropical oceans each year, familiar enough to be given names. Hurricane Gilbert that caused devastation on the island of Jamaica and the coast of Mexico in 1988, Typhoon Mireille that hit Japan in 1991, Hurricane Andrew that caused a great deal of damage in Florida and other regions of the southern United States in 1992 and Hurricane Mitch that caused great devastation in Honduras and other countries of central America in 1998 are notable recent examples. Low-lying areas such as Bangladesh are particularly vulnerable to the storm surges associated with tropical cyclones; the combined effect of intensely low atmospheric pressure, extremely strong winds and high tides causes a surge of water which can reach far inland.

In one of the worst such disasters in the twentieth century over 250 000 people were drowned in Bangladesh in 1970. The people of that country experienced another storm of similar proportions in 1999 as did the neighbouring Indian state of Orissa also in 1999, and smaller surges are a regular occurrence in that region. The increase in storm intensity during recent years has been tracked by the insurance industry, which has been hit hard by recent disasters.

Until the mid 1980s, it was widely thought that windstorms or hurricanes with insured losses exceeding one billion US dollars were only possible, if at all, in the United States. But the gales that hit western Europe in October 1987 heralded a series of windstorm disasters which make losses of ten billion dollars seem commonplace. Hurricane Andrew, for instance, left in its wake insured losses estimated at nearly twenty-one billion dollars with estimated total economic losses of nearly thirty-seven billion dollars.

It shows an increase in economic losses in such events by a factor of over 10 in real terms between the 1950s and the 1990s. Some of this increase can be attributed to the growth in population in particularly vulnerable areas and to other social or economic factors; the world community has undoubtedly become more vulnerable to disasters. However, a significant part of it has also arisen from the increased storminess in the late 1980s and 1990s compared with the 1950s. Windstorms or hurricanes are by no means the only weather and climate extremes that cause disasters.

Floods due to unusually intense or prolonged rainfall or droughts because of long periods of reduced rainfall can be even more devastating to human life and property. These events occur frequently in many parts of the world especially in the tropics and sub-tropics. There have been notable examples during the last two decades. In 1988, the highest flood levels ever recorded occurred in Bangladesh, eighty per cent of the entire country was affected; China experienced devastating floods affecting many millions of people in 1991, 1994-5 and 1998; in 1993, flood waters rose to levels higher than ever recorded in the region of the

Mississippi and Missouri rivers in the United States, flooding an area equivalent in size to one of the Great Lakes; major floods in Venezuela in 1999 led to a large landslide and left 30 000 people dead, and two widespread floods in Mozambique occurred within a year in 2000-1 leaving over half a million homeless.

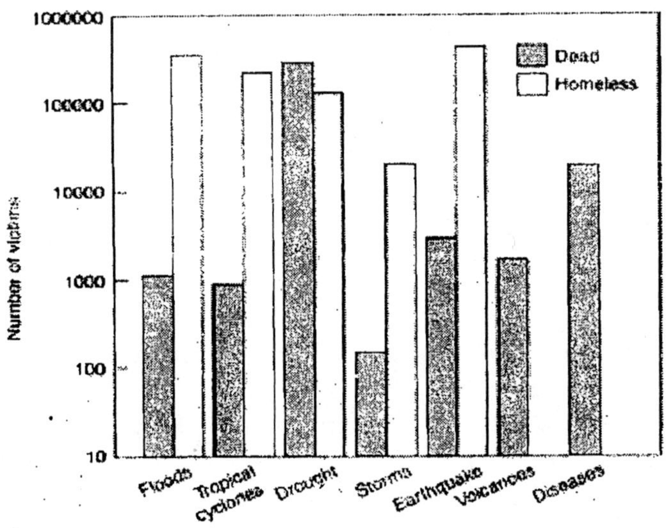

Figure 1. Recorded disasters in Africa, 1980–9, estimated by the Organisation for African Unity.

Droughts during these years have been particularly intense and prolonged in areas of Africa, both north and south. It is in Africa especially that they bear on the most vulnerable in the world, who have little resilience to major disasters. Figure 1 shows that in the 1980s droughts accounted for more deaths in Africa than all other disasters added together and illustrates the scale of the problem.

El Niño Events

Rainfall patterns which lead to floods and droughts

especially in tropical and semi-tropical areas are strongly influenced by the surface temperature of the oceans around the world, particularly the pattern of ocean surface temperature in the Pacific off the coast of South America. About every three to five years a large area of warmer water appears and persists for a year or more. Because they usually occur around Christmas these are known as El Niño events. They have been well known for centuries to the countries along the coast of South America because of their devastating effect on the fishing industry; the warm top waters of the ocean prevent the nutrients from lower, colder levels required by the fish from reaching the surface.

A particularly intense El Niño, the second most intense in the twentieth century, occurred in 1982-3; the anomalous highs in ocean surface temperature compared to the average reached 7°C. Droughts and floods somewhere in almost all the continents were associated with that El Niño. Like many events associated with weather and climate, El Niño often differ very much in their detailed character; that has been particularly the case with the El Niño events of the 1990s. For instance, the El Niño event that began in 1990 and reached maturity early in 1992, apart from some weakening in mid 1992, continued to be dominated by the warm phase until 1995.

The exceptional floods in the central United States and in the Andes, and droughts in Australia and Africa are probably linked with this unusually protracted El Niño. This, the longest El Niño of the twentieth century, was followed in 1997-8 by the century's most intense El Niño that brought exceptional floods to China and to the Indian sub-continent and drought to Indonesia—that in turn brought extensive forest fires creating an exceptional blanket of thick smog that was experienced over a thousand miles away. A scientific question that is being urgently addressed is the possible link between the character and intensity of El Niño events and global warming due to human-induced climate change.

Effect of Volcanoes on Temperature

Volcanoes inject enormous quantities of dust and gases into the upper atmosphere. Large amounts of sulphur dioxide are included, which through photochemical reactions using the Sun's energy are transformed to sulphuric acid and sulphate particles. Typically these particles remain in the stratosphere for several years before they fall into the lower atmosphere and are quickly washed out by rainfall. During this period they disperse around the whole globe and cut out some of the radiation from the Sun, thus tending to cool the lower atmosphere.

One of the largest volcanic eruptions in the twentieth century was that from Mount Pinatubo in the Philippines on 12 June 1991 which injected about twenty million tonnes of sulphur dioxide into the stratosphere together with enormous amounts of dust. This stratospheric dust caused spectacular sunsets around the world for many months following the eruption. The amount of radiation from the Sun reaching the lower atmosphere fell by about two per cent. Global average temperatures lower by about a quarter of a degree Celsius were experienced for the following two years. There is also evidence that some of the unusual weather patterns of 1991 and 1992, for instance unusually cold winters in the Middle East and mild winters in western Europe, were linked with effects of the volcanic dust.

Over the centuries different human communities have adapted to their particular climate; any large change to the average climate tends to bring stress of one kind or another. It is particularly the extreme climate events and climate disasters which emphasise the importance of climate to our lives and which demonstrate to countries around the world their vulnerability to climate change—a vulnerability which is enhanced by rapidly increasing demands on resources.

The range of normal natural climate variation is large. Climate extremes are nothing new. Climate records are continually being broken. In fact, a month without a broken record somewhere would itself be something of a record! Changes in climate that indicate a genuine long-term trend can only be identified after many years. However, we know for sure that, because of human activities especially the burning of fossil fuels, carbon dioxide in the atmosphere has been increasing over the past two hundred years and more substantially over the past fifty years.

To identify climate change related to this carbon dioxide increase, we need to look for trends in global warming over similar lengths of time. They are long compared with both the memories of a generation and the period for which accurate and detailed records exist. Although, therefore, it can be ascertained that there was more storminess, for instance, in the region of the north Atlantic during the 1980s and 1990s than in the previous three decades, it is difficult to know just how exceptional those decades were compared with other periods in previous centuries.

There is even more difficulty in tracking detailed climate trends in many other parts of the world, owing to the lack of adequate records; further, trends in the frequency of rare events are not easy to detect. The generally cold period worldwide during the 1960s and early 1970s caused speculation that the world was heading for an ice age. During the last few years, as the occurrence

2

Nature of Global Climate Change

Environmental change is a fact of life on earth. The predicted rapid changes in climate over the next century may be such that the subsequent changes in habitat and in species composition and diversity will result in ecosystems that function differently from the way they do now. For simplicity in thinking about climate changes, we can divide the earth's systems into the atmosphere (air), hydrosphere (water, predominantly the oceans), biosphere (life), geosphere (land, whether above or below the ocean surface), and cryosphere (snow and ice). It is important to realise that climate is affected by interactions within and among all these spheres.

Climate changes reflect variations within the Earth's environment, natural processes going on around it, and the impact of humans. The external factors which can shape climate are often called climate forcings and include such processes as variations in solar radiation, the Earth's orbit, and greenhouse gas concentrations.

Environmental Features

Meteorological Features

Temperature

The Arctic receives less heat energy as solar radiation than it emits into space as longwave radiation. This net loss of energy produces low temperatures. It is balanced by a redistribution of heat from southern latitudes through air and ocean currents. Topography and the distribution of land and sea create significant temperature differences at similar latitudes. For example, at a station in the Canadian Arctic Archipelago, the average January temperature is approximately 20°C lower than the temperature at a station at the same latitude on Svalbard.

Climate is divided into maritime and continental subtypes. Maritime areas have moderately cold, stormy winters. Summers are cloudy, with mean temperatures of about 10°C. Continental climate is found in the interior of the larger land masses in the Arctic, where precipitation is lower than over maritime areas, and temperature differences are more significant (e.g., July mean temperatures of +5°C to +10°C; January mean temperatures of –20°C to –40°C).

Precipitation

On average, the annual precipitation in the Arctic is less than 500 mm, typically between 200 and 400 mm. Cold air can hold less moisture than warmer air, and although the frequency of precipitation may be high, the overall intensity is low. This explains why the accumulation of snow is relatively low in winter in much of the Arctic.

Cloud cover

Persistent and extensive cloud cover is an important feature over polar oceans. The cloud cover varies from 70 to 90 per cent in summer to 40 to 60 per cent in winter. During periods of cold air outbreaks from the Arctic Basin, clouds are formed by cooling of the warmer boundary layer air.

Fog

Another characteristic feature of Arctic weather is fog. Some areas have, on average, more than 100 foggy days per year. In summer, the ice retreats northward, exposing open water, and warm air moves in over the ice and cold water, resulting in fog. In winter "sea smoke" or steam fog forms over open water leads in the pack ice.

Winds

The generally open landscape of the Arctic region means that winds are not greatly slowed by friction at ground level. Wind is an important factor in snow distribution, causing scouring in exposed areas and deposition in sheltered locations. It also augments the chilling effect of low temperatures. Wind also affects sea surface stability, increases mixing in the water column, and it influences ice drift.

Arctic Landscape

Fresh Water

The Arctic landscape is dominated by fresh water in the form of rivers, lakes, glaciers, permafrost, and wetlands. Glaciers in Greenland and other parts of North America contain a noteworthy proportion of the world's fresh water. Cold winters convert most fresh water to ice, which has had a significant impact on the shape of the landscape. Glacier-formed valleys and moraines, outwash plains, polygon tundra, and pingos, as well as river and coastal erosion, all result from endless seasonal cycles of ice freeze and thaw.

Arctic Rivers

Each year, Arctic rivers carry about 4200 km^3 of water into the Arctic Ocean along with a calculated 221 million tons of sediments. The landscape determines how much sediment each river carries. Those crossing flat, frozen tundra generally carry small amounts of sediment. Fast-moving rivers that flow through landscapes with less permafrost and with soils that are easily eroded, Canada's Mackenzie River for example, carries large amounts of sediment.

Permafrost

When ground is frozen year round, for more than two years, a typical feature of the Arctic, it is called permafrost. Permafrost may be anything from a couple of metres to one kilometre deep. Normally, only the uppermost, active layer (down to one metre) thaws during summer. The permafrost in northern Siberia and northern North America is deep and continuous.

Further south the permafrost is discontinuous and restricted to mountain areas. Permafrost stores large amounts of ancient carbon and methane, both global warming gases. Thawing permafrost is likely to release some of this stored carbon and methane back into the atmosphere, increasing the risk of further climate change.

The Arctic Ocean

The Arctic Ocean is the smallest of the world's five oceans. Most of the water flowing into the Arctic Ocean comes from the Atlantic Ocean through the North Greenland Sea and the Barents Sea. Each year, approximately two per cent of its water comes from rivers of the Arctic region. Nearly half of the ocean floor is continental shelf, the highest percentage among all oceans. These shelves are the primary areas of ice formation and ice melt.

The Arctic Ocean features an enormous ice pack of some 8 million km^2 in late summer, which nearly doubles in size at its peak between March and May. The ice pack is surrounded by open seas during the summer months. The ocean is almost entirely ice locked from October to June. The Arctic Ocean ice is not a monolithic cap, but moves, changes, breaks up, and reforms. It has an average thickness of approximately three metres with pressure ridges as much as three to five times thicker.

Ocean Currents

As in the atmosphere, there are both annual and inter-annual variabilities in ocean temperature, most pronounced in the relatively warmer water masses. In the North Atlantic, the ocean appears to alternate between warm and cold states, with fluctuation periods of three to five years as the most frequent.

Climate Change

Greenhouse Effect

The temperature of the earth is controlled by the presence of water vapour, carbon dioxide (CO_2), ozone, methane, nitrous oxide, and other naturally occurring atmosphericgases. These gases block outgoing, long-wave (heat) radiation from leaving the earth, thus warming the atmosphere (Figure 1). This effect has given birth to the common name greenhouse gases. On average, about one third of the solar radiation that hits the earth is reflected back to space.

Of the remainder, some is absorbed by the atmosphere but most is absorbed by the land and oceans. The earth's surface becomes warm and, as a result, emits infrared radiation. Greenhouse gases mixed throughout the atmosphere trap the infrared radiation, thus warming the atmosphere. Naturally occurring greenhouse gases create a natural greenhouse effect. However, human activities are

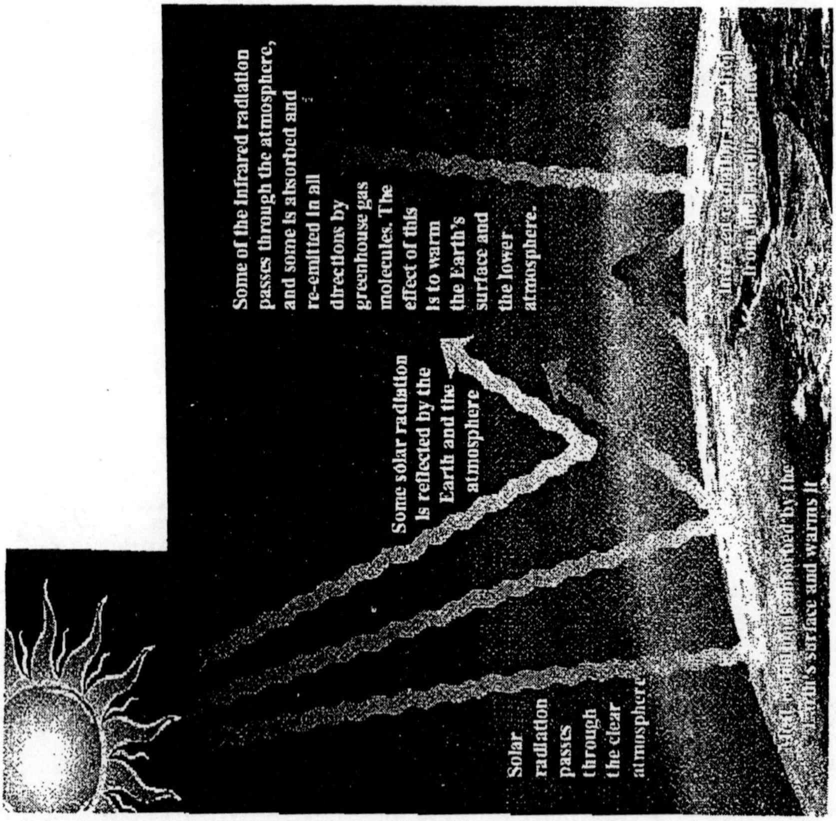

Figure 1. The Greenhouse Effect

causing greenhouse gas levels in the atmosphere to increase.

Cooling Effects

The amount of aerosols (suspended particles) in the air has a direct effect on the amount of solar radiation that can reach the earth's surface. Aerosols may have significant local or regional impact on temperatures.

Changes in Atmospheric Carbon Dioxide

Concerns about climate change stem from the increasing concentration of CO_2 and other greenhouse gases). Atmospheric CO_2 has increased from a pre-industrial concentration of about 280 ppmv (parts per million by volume) to about 367 ppmv at present. The rapid increase has been occurring since the onset of industrialisation, and has closely followed the increase in CO_2 emissions from burning fossil fuels.

Changes in Surface Air Temperature

Surface air temperatures seem to be increasing by about 1.5°C per decade over continental central Siberia and over continental North America. In Fennoscandia, the records do not indicate any significant changes, whereas Baffin Bay is cooling by 1.5°C per decade. North of 70°N, temperature observations are sparse. There are indications, however, of warming around the northern continental rims of central and western North America and central Asia over the past century.

Increasing air temperatures are evident at both sea and land surfaces. Indirect indicators, such as borehole temperatures and glacier shrinkage, provide independent support for the observed warming. It should be noted that the warming has not been globally uniform. For eastern

North America through the North Atlantic, there is a cooling trend.

Changes in Precipitation

During the past 40 years, precipitation has increased by up to 15 percent in high latitudes. On the North American tundra, there is a trend toward earlier spring snowmelt. Changes in snow cover may have profound impacts on plant and animal life in the Arctic.

Changes in Permafrost

Much of the permafrost in the Arctic is close to 0°C and particularly sensitive to temperature changes. Measuring temperature profiles in permafrost can provide a climate record that goes back hundreds of years at a particular site. Such measurements from Alaska indicate a warming of 2°C to 4°C over the past hundred years. This warming has been confirmed by observations that the discontinuous permafrost in Alaska is thawing.

Inuit in Barrow, Alaska, have seen their ice cellars, which are dug into the permafrost, drip water for the first time in living memory. Widespread loss of discontinuous permafrost will trigger erosion or subsidence of ice-rich landscapes, and change hydrologic processes. Large-scale thawing of permafrost will make the Arctic tundra into a major source of carbon dioxide and methane-two important greenhouse gases. In addition, changes in the cryosphere will reduce slope stability and increase incidence of natural hazards for people, structures, and communication links. Buildings, roads, pipelines, and communication links will be threatened or destroyed.

Sea Ice

Sea ice plays a critical role in the energy budget of the Arctic and thus in the region's climate. Snow-covered ice is highly reflective, so decreasing sea-ice cover will enhance a warming trend. Sea ice is also a physical barrier between the ocean and the atmosphere. Less ice and warmer air would allow the air to pick up moisture from the water, which might make the Arctic cloudier. This would probably change regional weather patterns. Sea ice limits the exchange of carbon dioxide between water and air and the penetration of light into the water, which in turn affects the productivity of algae in the ocean.

The thickness of the part of the ice that is submerged is its draft. Comparison of sea-ice draft data acquired on submarine cruises between 1993 and 1997 with similar data acquired between 1958 and 1976 indicates that the mean ice draft at the end of the melt season has decreased by about 1.3 m in most of the deep water portion of the Arctic Ocean, from 3.1 m in the period 1958 to 1976 to 1.8 m in the 1990s.

Rising Sea Levels

According to IPCC, the sea level will rise somewhere between 15 and 95 centimetres by 2100. In the Arctic, many smaller glaciers have been shrinking during the past century and global warming will accelerate their demise. The mass balance of glaciers is controlled not only by temperature but also by snowfall and the physical processes of ice motion. Therefore, changes in glacier size typically lag behind climate changes, by years to decades for mountain glaciers and by longer periods for larger ice sheets.

Water Temperatures

Sea levels will rise further because water expands when it warms up. Moreover, the huge amount of fresh water from melting glaciers is likely to affect seawater salinity, which will, in turn, affect the mixing of water masses and ocean currents. Changes in the formation of sea ice, which leads to salty, heavy water sinking down deep into the ocean, also plays a key role in driving deepwater currents, which, in turn, affect weather patterns and climate around the globe. Open water during part of the year over the shallow continental shelf of the Arctic Ocean will also allow exchange of contaminants between water and air, and high biological productivity will provide routes for contaminants to enter the food chain.

POLLUTANTS

Arctic Haze

The term Arctic haze was coined in the 1950s to describe an unusual reduction in visibility observed in the High Arctic. The haze is seasonal, with a peak in the spring. It is caused by human activity; the main source are sulphate emissions from coal burning, mostly in Eurasian countries.

Impacts of Chemicals

Due to global atmospheric and ocean transportation patterns, many chemicals, released into air or water by human activities, may accumulate in the Arctic in concentrations that have ecological impacts on humans, fauna and flora. Damages, to human beings and animals, from hazardous substances may include physical genetic defects, metabolic changes, reduced fertility, and cancer. Nervous system and muscle functions may also be affected.

These impacts could lead to loss of habitat and may be detrimental to the health and welfare of entire populations. Examples of such hazardous substances are heavy metals (e.g., mercury), persistent organic pollutants (POPs) and radioactive materials. Several of these substances can accumulate in food chains until toxic levels are reached. Top predators are particularly vulnerable to such bioaccumulation. Combined effects of various environmental contaminants have produced unforeseen problems.

Mercury Level

The levels of mercury in peat bogs in southern Greenland have increased tenfold during the 20th century, reflecting changes in atmospheric deposition. The pattern of decreasing heavy metal emissions in Europe and North America has been offset by increased emissions associated with the growing demand for electricity in Asia. Globally, emissions of heavy metals from stationary fossil fuel combustion have not changed significantly since the beginning of the 1980s.

Recent studies have identified a mechanism by which elemental mercury, previously thought to be of little biological significance, may be converted to a reactive form that accumulates in biota in the Arctic (AMAP 2003). On the Faeroe Islands, subtle effects have been observed on infant neurodevelopment and blood pressure due to exposure to mercury during pregnancy. The level of mercury observed in other populations implies that infants may be at risk of similar effects in several parts of the Arctic.

The Arctic Council has identified mercury as an emerging issue and has requested UNEP to initiate a global assessment that could form the basis for appropriate international action. Polar bears in northern Alaska have higher levels of mercury in liver and muscle tissue than do those in western Alaska. Bears in western Arctic Canada seem to accumulate mercury faster in their livers than polar bears in eastern Arctic Canada. The eastward decreasing trend in mercury extends to Greenland and Svalbard.

The geographic difference found between western Arctic Canada and eastern Arctic Canada is most likely caused by higher level of mercury levels in the ringed seal food chain caused by higher natural levels in the sediments-and consequently in the lower food chain of the Mel-ville Island area. Mercury acts as a nerve toxin and the main health concern is its effect on the brain, particularly in the growing foetus and the young. Mercury can also damage reproduction in mammals by interfering with the formation of sperm. Neurological and reproductive effects have also been seen in birds. In fish, its effects include a decreased sense of smell, damage to the gills, blindness, and more.

Persistent Organic Pollutants (POPs)

The Arctic accumulates pollutants, including POPs like polychlorinated biphenols (PCBs) and the pesticide DDT that have been discharged from industry and agriculture around the globe. Large amounts of these "ghosts of the past" that are trapped in sea ice could be released and allowed to enter the food chain through atmospheric warming. This would pose a health threat to top predators such as humans and polar bears.

Radioactive Pollutants

Strontium and cesium fallout from the Chernobyl accident and past nuclear tests-accumulates in lichen, which is, in turn, eaten by reindeer and caribou, and then by humans. As a former playground of the Cold War, the Arctic faces the risk of radioactive contamination from leakage from old military installations, weapon storage sites, reprocessing plants, and ships carrying nuclear waste from Europe to central Siberia.

Acidification

Oxides of sulphur are the major acidifying compounds in

the Arctic. They are formed when fossil fuels burn and when sulphide ores are smelted. Release of sulphur dioxide (SO_2) to the environment has risen with the growth in energy demand and industrial activity. Most of the sulphur in Arctic air comes from industrial areas further south. Eurasia (40 per cent) and eastern North America (20 per cent) are the major global sources. A large part of the remaining global emissions occur in the Far East, particularly China.

Emissions of SO_2 have decreased considerably in North America and Europe after peaking in the late 1970s and early 1980s. In the Arctic, production of copper, nickel, and other non-ferrous metals from sulphur-bearing ores creates the largest emissions of acidifying substances. Most smelter emissions come from the Nikel, Zapolyarnyy, and Monchegorsk complexes on the Kola Peninsula and from Noril'sk in northwestern Siberia. Compared with similar industries in other areas, emissions from these smelters are extremely high.

Environmental Impacts

While it is one of the world's last, large, mainly undisturbed terrestrial ecosystems, the Arctic now faces problems due to climate change, ozone depletion, long-range transported pollution and unsustainable use. These anthropogenic risks loom as the greatest threats to the biodiversity of the Arctic. Seals, caribou, fish, whales, and polar bears may also pass along high levels of contaminants to human communities that subsist on marine mammals and wildlife.

The most biologically productive and diverse areas in the Arctic are along coasts and at the edges of the ice, where marine mammals are most abundance. These areas are some of the largest productive marine ecosystems on earth, and have, for this very reason, faced strong pressure from fisheries and hunting of marine mammals over the last decades.

The Bering and Barents seas currently support the world's largest fisheries by tonnage. Shore birds, caribou, whales, polar bears, seals, and sea lions all can be found in abundance at the long polar coastline. These ecosystems are vulnerable to rapid changes, though there is a record of hardy species that can quickly recover.

For the marine wildlife that depends on these areas for breeding and hunting-and Indigenous communities that still maintain a subsistence lifestyle-the earlier breakup, steady retreat, and later return of shorefast ice present critical challenges to survival. Another threat to Arctic biodiversity is habitat fragmentation. While the Arctic does include some of the largest tracts of wilderness on the planet, infrastructure development, mainly due to presence of minerals and hydrocarbons, has, for decades, influenced even remote corners of the North. Thus, infrastructure development can be used as an indicator of human impacts in the Arctic.

Influence of Humans

While Indigenous human habitation in northern latitudes dates back several thousand years in most areas of the Arctic, those settlements have remained modest. Although non-Indigenous residents now outnumber Indigenous residents in much of the Arctic, Indigenous communities continue to redefine their relationship between living on the land using traditional means and adopting and taking advantage of modern approaches and technology. Steadily growing populations throughout the twentieth century attest to the continued vibrancy of Indigenous peoples of the North.

Dictated by topography and harsh climate, employment in the Arctic is mainly a mixture of nomadic animal husbandry, tourism, and natural-resource extraction. In the twentieth century, the Russian North has experienced a mass immigration due to resource extraction industries-coal, oil, gas, nickel, copper, gold, and diamonds. In a vast area with approximately 180,000 Indigenous people, over 6 million ethnic Russians have settled in northern urban centres. Some of these centres have from one- quarter- million to one million residents.

Since the demise of the Soviet Union, however, that influx has begun to reverse en masse. The survival of Arctic peoples has always been intricately linked to climate. To the extent that people continue to harvest plants and animals and live permanently or seasonally on coastal spits or along riverbanks and lakeshores, climate change will directly impact their lives. Even if modern resources can mitigate some effects, climate change is likely to disrupt culturally important hunting and fishing activities.

Coastal erosion has forced some Aboriginal communities in Alaska to relocate. A rise in sea level would threaten many more communities, especially in Russia and Alaska where they are often located on low-lying coastal plains and on river deltas. Coastal erosion would also cause changes to the geography of river deltas. Temperature and humidity changes will probably affect the local physical environment. Thawing permafrost represents a growing challenge with huge economic implications.

On large areas of the coastal plains and wet tundra, an increase in precipitation could contribute to making the land unusable. It might also shift the migration patterns of terrestrial mammals and alter the breeding and molting areas of birds. Changes in snow cover would alter travelling conditions over the tundra, making it potentially more difficult for hunters to reach inland locations in spring and fall. For communities on the coast, changes in sea ice might have dramatic impacts by shifting the migration routes of marine mammals.

Even if animal life seems abundant, it is often made up of seasonal migrants on their way to specific feeding grounds where food production is intense but brief. The population density of seals, for example, is correlated with the distribution of coastal sea ice. Indigenous peoples and other northern residents depending on traditional food for all or part of their diet may be adversely affected by chronic exposure to long-distance pollutants.

While fragmented over three continents and two islands- North America, Europe, Asia, Greenland and Iceland-problems and concerns confronting the region in many ways share more in common across the Circumpolar North than between northern and southern regions of the same continent or nation state. Since the end of the Cold War, this sense of common history and purpose has led to a growing number of post-Cold-War avenues for Arctic cooperation. Among all collaborative efforts needed for sustainable development in the region, concerted action on environment and climate change issues will require top priority.

3

Global Warming: Causes and Consequences

Global warming, or for that matter any substantial warming of the Earth's surface, begins with the sun. Except for relatively small fluctuations due to sunspot activity, the amount of radiation from the sun that reaches the Earth has been fairly constant from year to year and century to century. If you were to travel to the outer reaches of the atmosphere and hold up a flat surface perpendicular to the sun's rays for several years during the daylight hours, you'd find that about 1,368 Watts per square meter on average would hit that surface. Not all of that energy would be absorbed by the Earth.

Roughly 30 percent of the total solar energy that strikes the Earth is reflected back into space by clouds, atmospheric aerosols, reflective ground surfaces, and even ocean surf. The remaining 70 percent is absorbed by the land, air, and the oceans. The absorbed light is mostly in the form of ultraviolet, visible, and near-infrared solar radiation. Absorption of solar energy heats up our planet's surface and atmosphere and makes life on Earth possible. The energy does not stay bound up in the Earth's environment forever. If it did, then the Earth would grow hotter and hotter until its temperature exceeded that of the sun. Instead, as the rocks, the air, and the sea heat, they emit thermal radiation.

Much of this thermal radiation, which is largely in the form of longwave infrared energy, travels directly out into space, leaving the Earth and allowing it to cool. Such radiation is invisible to our eyes, but our hands can feel it radiating from a fire or a car engine. Some of this outgoing longwave infrared radiation, however, is re-absorbed by water vapour, carbon dioxide, and other greenhouse gases in the atmosphere and is then re-radiated back toward the Earth's surface. On the whole this re-absorption process is good. If there were no greenhouse gases or clouds in the atmosphere, the Earth's average surface temperature would be a very chilly –18°C instead of the comfortable 15°C that it is today.

What has many people worried now is that over the past 250 years humans have been artificially raising the concentration of greenhouse gases in the atmosphere. Our factories, power plants, and cars burn coal and gasoline and spit out a seemingly endless stream of carbon dioxide. We produce millions of pounds of methane by allowing our trash to decompose in landfills and by breeding large herds of methane-belching cattle. Nitrogen-based fertilizers, which we use on nearly all our crops, release unnatural amounts of nitrogen oxide into the atmosphere. Once these carbon-based greenhouse gases get into the atmosphere, they stay there for decades or longer.

According to the Intergovernmental Panel on Climate Change (IPCC), since the industrial revolution, carbon dioxide levels have increased 31 percent and methane levels have increased 151 percent. Paleoclimate readings taken from ice cores and fossil records have shown that these gases, two of the most abundant greenhouse gases, are at their highest levels in the past 420,000 years. Many fear that the increased concentrations of greenhouse gases have prevented additional thermal radiation from leaving the Earth.

In essence, these gases are trapping excess heat in the Earth's atmosphere in much the same way that a windshield traps solar energy that enters a car. Much of the available climate data appears to back these fears. Temperature data gathered from many different sources all across the globe have shown that over the last 100 years the surface temperature of the Earth, which includes the lower atmosphere and the surface of the ocean, has risen dramatically over the past century.

The IPCC estimates the increase has been between 0.4°C and 0.8°C. Worldwide measurements of sea level have shown a rise of 0.1 to 0.2 meters over the last century. Readings gathered from glaciers reveal a steady recession of the world's continental glaciers. Taken together, all of these data suggest that over the last century the planet has experienced the largest increase in surface temperature in 1,000 years. As of now, greenhouse gases afford a plausible explanation for such changes. Paleoclimate readings back this hypothesis up.

In the Earth's past, drastic increases in carbon dioxide nearly always coincide with large increases in Earth surface temperatures. Conversely, ice ages are almost always accompanied by a decrease in carbon dioxide. Logic dictates that, as third world nations develop their economies and first world nations consume more energy, greenhouse gas concentrations will continue to rise. Though scientists have not reached a consensus, most leading researchers and organisations purport that the average surface temperature of the Earth will increase along with increasing emissions. According to the IPCC, the surface temperature could rise by between 1.4°C and

5.8°C by the end of the century. Scientists at Goddard Institute for Space Studies, NASA's division spearheading climate modelling efforts, report that we should expect between 0.5°C and 1°C over the next fifty years.

At first glance, these numbers probably do not seem threatening. After all, temperatures typically change a few degrees whenever a storm front moves through. Such temperature changes, however, represent day-to-day regional fluctuations. When surface temperatures are averaged over the entire globe for extended periods of time, it turns out that the average is remarkably stable. Rarely in the Earth's history has the average surface temperature changed as dramatically as the changes that scientists are predicting for the next century.

During the last ice age 20,000 years ago, for instance, the Earth was roughly 5°C cooler than it is today. Since then it has warmed up, although not steadily, to present levels. That's an increase of roughly 1°C every 4,000 years. Current global warming scenarios predict this bare minimum increase over the next century.

Influence of Global Warming

How all of this warming will alter the weather is more uncertain. It's much easier for scientists to forecast the Earth's average surface temperature than it is to forecast how much rain will fall in, say, Boise, Idaho, during the next in 50 years. So far scientists have not been able to pinpoint with certainty any changes in weather due to global warming over the last century. Most of that "weird" weather we've been experiencing-that unusually warm fall or that particularly wet winter-is due to normal, regional changes in the weather.

Some scientists believe that global warming will continue to have relatively little impact on the day-to-day climate conditions. Others purport that future changes will likely be subtle, and they will spread over large areas of the globe from decade to decade and creep up on us like old age. Still others hypothesize that when the Earth's surface temperature reaches some critical threshold, the heat will trigger relatively drastic changes to the atmosphere and the oceans and transform the Earth's weather patterns in a matter of years. Not surprisingly, many scientists speculate that such changes in the climate will probably result in more hot days and fewer cool days.

According to the IPCC, land surface areas will increase in temperature over the summer months much more than the ocean. The mid-latitude to highlatitude regions in the Northern Hemisphere- areas such as the Continental United States, Canada, and Siberia-will likely warm the most. These regions could exceed mean global warming by as much as 40 percent. Forecasts for precipitation and weather are cloudier.

Right now the IPCC reports that the amount of precipitation, especially in the mid-latitude to high-latitude regions of the Northern Hemisphere, will likely increase. They believe, however, that it will come in the form of bigger, wetter storms, rather than in the form of more rainy days. So it's more probable that the increase in rain will only serve to tax our drainage systems rather than benefit vegetation or replenish natural, underground aquifers. As to larger more destructive weather patterns, hurricanes will likely increase in intensity due to warmer ocean surface temperatures. The outlook for rising sea levels is nothing like the deluge portrayed in Hollywood.

The statue of liberty won't be up to her neck in water, and we won't all be living on flotillas on an end-less sea. According to the IPCC, over the next century, sea levels are likely to rise between 0.09 and 0.88 meters. The rise will mainly be due to seawater expanding from the increased ocean temperatures and run-off from the melting of continental glaciers and a slight melting of the Greenland Ice Sheet. For now, the West Antarctic Ice Sheet, which could raise our sea levels dramatically, will probably stay in place. It may even gain more mass due to an increase in precipitation over the next century. But, if somehow the entire Greenland Ice Sheet melted and the West Antarctic Ice Sheet fell into the sea, the sea level would rise roughly 10 meters. This is probably impossible over the next century, but there is the danger that global warming could initiate ice sheet changes that will continue to develop over future centuries.

Should global warming continue, many biologists envision the alteration of natural habitats. Some of this change may be for the better. Higher levels of CO_2 and warmer temperatures may cause forests to become more lush and vigorous. Warmer ocean waters on the open ocean could be beneficial to fish and algae on the high seas. Unfortunately, most changes will likely be for the worst. Plants and animals in mid-latitude regions, such as nut-bearing oaks in the midwestern United States, may find themselves in warmer environments where they cannot survive. Rising sea levels may inundate delicate coastal wetlands with brackish waters, which could drive out certain types of fish and kill wetland vegetation.

Warmer ocean temperatures around the coast could overheat many types of coral, killing them and many of the animals that depend on them. As far as human health is concerned, those hit hardest will probably be residents of poorer countries that do not have the funds to fend against changes in climate. A slight increase in heat and rain in equatorial regions would likely spark an increase in vector-borne diseases such as malaria. More intense rains and hurricanes could cause more severe flooding and more deaths in coastal regions and along riverbeds. Even a moderate rise in sea level could threaten the coastlines of lowlying islands such as the Maldives.

All across the globe, hotter summers could lead to more cases of heat stroke and deaths among those who are vulnerable, such as older people with heart problems. The warmer temperatures may also lead to higher levels of near-surface ozone from cars and factories, which would likely cause more code red air quality days and hospital admissions for those with respiratory problems.

The severity of these environmental changes will be largely dependent on how much the Earth's surface warms over the next century. As the wide range of estimates for average global surface temperature suggests, researchers haven't exactly reached a consensus. The reason for the wide range simply comes down to the difficulty inherent in predicting the outcomes of current trends in both human society and the Earth's climate system. To get their estimates for future warming, scientists must first discern how much human industry and expansion will impact the Earth over the next 100 years. Researchers typically review a wide range of socioeconomic data such as worldwide population trends and then come up with varying scenarios describing mankind's future.

The worst case scenarios, typically known as the "business as usual" scenarios, go on the assumption that population growth in developing nations will proceed unabated and the entire world will continue to use more and more fossil fuel per capita. The best case scenarios envision a world in which environmentally friendly technologies such as fuel cells and solar panels replace much of today's fossil fuel combustion and the population increase is halted by an improved standard of living worldwide. Using these projections, researchers arrive at estimates for how much carbon dioxide, soot, ozone and other pollutants people will put into the air over the next 100 years.

Obviously, the "business as usual" scenarios have people producing the most pollution, and the eco-friendly scenarios have people producing the least pollution. But projections of greenhouse gas concentrations alone cannot tell scientists how much the Earth's surface will warm or the climate will change. To make forecasts, they must employ climate models, which are essentially computer simulations of the climate. These models are a bit like those computer programmes detectives use to envision what missing persons would look like ten years after their disappearance. But instead of being constructed on the knowledge and data on how people's faces age, these models are constructed on the knowledge and data of the Earth's climate.

After inputting estimates for future greenhouse gas emissions, scientists run the models forward into many possible futures. Some of these models simply produce forecasts of the Earth's temperature, while others are built to predict other changes in the Earth's atmosphere and oceans. Even the simplest of these models can be exceedingly complex. When piecing together models to estimate the Earth's average surface temperature, scientists must take into account everything on the Earth that blocks outgoing thermal radiation or reflects sunlight into space as well as possible changes in the radiation emitted by the sun itself.

There are, in fact, a myriad of unnatural and natural factors in addition to greenhouse gases that could sway global surface temperatures one way or the other in the future. Some of these influences are human made and some are natural. Some directly impact the amount of radiation absorbed by the Earth, and scientists expect others will be triggered as the climate heats up. One example of a cooling influence that may counter the greenhouse effect is sulphate aerosols. When fossil fuels burn, they not only release greenhouse gases, but also sulphur dioxide.

The sulphur dioxide gets into the air and mixes with oxygen to create sulphate aerosol particles, which reflect sunlight. Though these aerosols don't stay in the air as long as most greenhouse gases, they may cool regions of the wcrld where their emissions are heavy, such as the eastern United States, China, and the Indian sub-continent. Changes in clouds due to the warming of the atmosphere, on the other hand, could greatly add to the warming effect. As the Earth heats up, the cloud composition in the atmosphere could change dramatically.

Low lying clouds could evaporate during the daytime hours and more high flying cirrus clouds could form as surface heat causes the air to rise. Since low lying clouds tend to reflect sunlight and higher clouds absorb heat radiation, the overall effect of these changes would be an increase in the amount of energy trapped in the atmosphere. On the other hand, the response of clouds to global warming is so uncertain that changes in clouds might act to reduce, not increase, the amount of surface warming over the next century. As if these warming and cooling influences weren't enough to keep track of, some evidence has shown global warming may have a delay built into it.

Given the level of carbon dioxide in the atmosphere now, a number of Earth scientists calculate that the Earth should be 0.5°C warmer than it is today. Temperature readings obtained from several hundred meters below the ocean's surface suggest that this extra energy could be lurking down there. The theory is that as the Earth has warmed, much of the excess energy has gone into heating the upper layers of the ocean, giving rise to a temperature imbalance between the surface waters and the deep waters.

It is thought that convection currents in the ocean may have transported some of this excess heat in the surface waters down deep, effectively removing it from the surface of our planet. Were this process to continue, the temperatures of the lower layers of the ocean would eventually increase until they are once again in balance with the surface waters. At this point, the excess heat in the upper layers would no longer be drawn down, and the Earth would warm to a higher level. So even if we drastically lower our emissions today, we could still be in for a 0.5°C additional warming.

Climate modelers must consider dozens of such factors, boil them down into equation form, and pack them into their models. Not all models are built alike. There is quite a lot of disagreement among Earth scientists as to how much of a role factors such as aerosols and clouds will play in heating the Earth and how they should be incorporated into the models. For instance, NASA climate modelers at GISS have evidence that black carbon aerosol particles (soot) contributes significantly to warming of the lower atmosphere, since they absorb incoming radiation. The IPCC, on the other hand, estimates that black soot plays only a very small role in warming. Each research group or agency builds its models accordingly, and the choices made influence the forecasts derived from the models. Even when modelers do agree on the mechanisms involved, many of these factors have a great deal of uncertainty associated with them.

The Skeptics

While the general consensus among scientists is that global

warming is real and its overall effects are detrimental, there are still some prominent scientists who feel that the threat of global warming has been greatly exaggerated. Skeptics take issue with the basic temperature data that demonstrate the Earth's temperature has increased over the last century. Most of the pre-satellite, pre-1970 data were collected in urban areas using many types of thermometers that were spread far apart. Such measurements are subject to human error and do not give a clear depiction of ocean temperatures. Until satellite data are collected for several more decades, some researchers feel that the temperature data remain too unreliable to take at face value.

Another point of contention is that no one has ever proven outside of the laboratory whether global warming occurs as a result of carbon dioxide. Scientists have ample fossil evidence that shows that carbon dioxide levels in the atmosphere have risen as the Earth grows warmer, but no one has yet shown that a rise in carbon is responsible for the past temperature increases. It is still possible that the warming in the distant past could have triggered the rise in carbon dioxide. As far as forecasts of future warming are concerned, skeptics point to the uncertainties inherent in the models researchers are using.

There are a couple of dozen models currently in use that forecast everything from the average surface warming of the planet to complex interactions global warming will have with the Earth's atmosphere and weather systems. As mentioned, each of these models can generate a different answer depending on projections for future human emissions, the uncertainty in how the climate will respond, and what scientists decide to include in the models. Many feel there is still too much we do not understand about the climate or human society to take stock in any forecasts as of yet.

But perhaps the one aspect of global warming research that nearly everyone agrees on is that more has to be done. NASA's Earth Science Enterprise is involved in a number of projects that aim to monitor and analyse climate change. A number of Earth Observing System (EOS) satellite missions have either already been launched or are slated to be launched over the next several years. Instruments aboard these satellites will take unprecedented measurements of the Earth and the sun that are relevant to climate change.

The EOS Terra satellite, which was launched in 1999, retrieves global readings of land surface temperatures, snow cover, atmospheric aerosol levels, cloud properties, methane, vegetation density, and a host of other variables that influence climate on a global scale. The measurements are taken over the entire globe once a day, often taken at resolutions of 1 km or less. Taking measurements by satellite is much more efficient, consistent, and timely than taking measurements in situ on the planet's surface. Since the early 1970s, researchers at GISS have been constructing climate models with the data and knowledge gathered in part from satellite readings.

They've constructed a number of radiative forcing models to try to understand the manner in which the Earth absorbs and reflects solar radiation. Such models could be used to make forecasts of the temperature of the Earth's surface and atmosphere. In addition, GISS researchers are working on several more comprehensive atmospheric and oceanic models in an effort to understand and forecast long-term changes in precipitation, air pressure, and wind currents. Were the scientists able to perfect these ocean atmosphere models, they could get a better handle on how rainfall and cloud cover will change if the Earth warms due to greenhouse gases.

Though data collection and modelling efforts have come a long way in the last 25 years, NASA and other agencies dealing with global warming are still a long way off from forecasting the future with certainty. But this new generation of satellite remote sensors are far more sophisticated than their predecessors. Their measurements are more precise and made more frequently over the entire globe every day. Moreover, they can measure a wider range of the Earth's vital signs, enabling scientists to better quantify the key cause-and-effect relationships that drive our climate. Each year dozens of discoveries are made that add to scientists' body of knowledge of how the climate works. Ultimately, the goal is to determine which climate changes are natural in origin, and which are due to human influences.

4

Role of Ocean in Climate Change

The ocean interacts with the atmosphere in two main ways. The first way is physically, through the exchange of heat, water, and momentum. Covering more than 70 percent of the Earth's surface and containing about 97 percent of its surface water, the ocean stores vast amounts of energy in the form of heat. Moreover, the ocean has a relatively large temperature inertia, or resistance to change.

Previously, scientists perceived the ocean as an unchanging "desert" due to its slow circulation and its low biological productivity. Yet, today we know the biological and physical functioning of the ocean system can change quickly over both small and large areas. Because it often drives the timing and patterns of climate change, the ocean was recently labelled by some scientists as the "global heat engine." As heat rises and eventually escapes the ocean to warm the atmosphere, it creates air temperature gradients and, consequently, winds. In turn, winds push horizontally against the sea surface and drive ocean current patterns.

Meanwhile, variations in temperature and salinity control vertical ocean currents-warmer, fresher waters flow upwards while colder, denser water tends to sink. Over time, a complex circulation pattern is established whereby warm surface waters move poleward where heat escapes more readily to outer space, while cold, deep currents are established in the ocean depths. Through this complex ocean circulatory system, the oceans and atmosphere work together to distribute heat and regulate climate. This circulation transports enormous amounts of heat, resulting in more moderate climates on land areas that are near the ocean. For example, London is warmer in winter than Toronto, even though London is closer to the North Pole.

Global climate modellers recently estimated that over the course of the 20th century, the ocean has reduced by about half the expected surface warming due to rising greenhouse gas levels. Scientists have observed an overall cooling trend in the east despite the strong and frequent El Niño events after 1975. Modellers conclude that as a consequence of the exchanges of heat and momentum between ocean and atmosphere, the mean temperature of the Pacific Ocean increases less than it would if it only exchanged heat. Thus, given its efficiency at redistributing heat poleward, the ocean is effectively delaying and regulating global warming. But scientists don't know if the ocean's role as "climate moderator" will persist over the long term.

There is evidence that large and abrupt changes in the ocean's circulation have had major impacts on global climate in the past. Some scientists question whether the existing ocean circulation pattern is stable enough to withstand the stresses of rising temperatures, heightened greenhouse gases and fresh water runoff due to melting glaciers. As the rate of fresh water runoff increases, it lessens the salinity and density of surface water, thereby raising its freezing point. More surface ice would inhibit the escape of heat and could result in a major reorganisation of the ocean's circulation system, which would, in turn, affect the global climate.

The second way that the ocean and atmosphere couple is chemical, as the ocean is both a source and sink of greenhouse gases. Much of the heat that escapes the ocean is in the form of evaporated water, the most abundant greenhouse gas on Earth. Yet, water vapour also contributes to the formation of clouds, which shade the surface and have a net cooling effect. In the long run, scientists don't know which process will exert the larger influence on global temperatures. Of the greenhouse gases, CO_2 is perhaps the most important because of its links with human activities.

Most of the world's carbon resides in the ocean, and the processes that result in exchanges between the surface ocean and the atmosphere, and between the upper ocean and the deep ocean, are critical. Carbonate chemistry regulates much of the transfer between these systems, but biological processes, such as photosynthesis which turns CO_2 into organic material, also play an important role. Settling of this organic carbon into the deep ocean is known as the "biological pump." In part, the upper ocean has lower concentrations of total carbon than the deep ocean as a result of the actions of this pump.

Some scientists speculate that if the ocean's circulation pattern is disrupted, it could become a source rather than a sink for carbon and atmospheric CO_2 levels could rise much higher than they are now. One of the key questions in climate change research is how the physical and biological processes of the ocean will respond to chemical and physical changes in the atmospheric. For example, will there be increased storms that increase mixing in the upper ocean? Will phytoplankton become more or less productive in such an environment? Windblown dust from soils and desert sands are rich in iron that, when they settle into the ocean, serve as "fertilizer" for the phytoplankton. Will this dust input increase as a result of climate change? Will this increase the efficiency of the biological pump and thereby increase the removal of CO_2 from the atmosphere by the ocean?

There is some evidence that larger amounts of iron may be deposited in the ocean in the future as global warming causes increases in windblown continental dust. Theoretically, if iron-containing aerosols fall in large regions of the ocean that are rich in other nutrients, then this would act like "fertilizer" to promote burgeoning phytoplankton populations, thus slowing the rate of CO_2 increase in the atmosphere and partially offsetting the anticipated warming. While we know that short-term events, such as El Niño, dramatically affect phytoplankton concentrations in the eastern Pacific, scientists aren't sure how, over the long-term, changes in ocean circulation and atmospheric forcing will affect the ocean's productivity.

During an El Niño event, the whole marine food chain is disrupted. Larger fish and mammals must either starve or move to where phytoplankton are more abundant. Linking these intense events with changes in climate variability is one of the most pressing issues in climate research. Yet, long-term trends are difficult to detect when natural variability is so high. A comprehensive programme of observations and modeling is necessary to improve our ability to make predictions on the future course of the Earth system.

Terra Spacecraft

Terra is the flagship of the Earth Observing System (EOS), a series of spacecraft to observe the Earth from the unique vantage point of space. Focused on key measurements identified by a consensus of U.S. and international scientists, EOS will enable research on the complex interactions of Earth's land, ocean, air, ice and life systems.

Terra will circle the Earth in an orbit that descends perpendicularly across the equator each day at 10:30 a.m. local time, when cloud cover is at a minimum and the space-based view of the surface is least obstructed. Each individual swath of measurements can be compiled into global images as frequently as every two days. Over a month or more, in combination with measurements from other polar orbiting satellites, Terra measurements will provide accurate monthly-mean climate assessments that can be compared with computer model simulations and predictions.

The Earth Observing System has three major components: the EOS spacecraft, an advanced ground-based computer network for processing, storing, and distributing the resulting data; and teams of scientists and applications specialists who will study the data and help users in industry, universities and the public apply it to issues ranging from agriculture to urban planning.

Space-based Oceanography

Reliable sea surface temperature measurements from space-based sensors have been a goal of oceanographers since the late 1960s. For the first time, Terra will provide oceanographers with the radiometric resolution and precision, the scientific calibration, the spatial coverage, and the ability to remove atmospheric effects that will enable measurements of sea surface temperature accurate to within 0.5°C. These data will enable a better understanding of ocean-atmosphere coupling-particularly during El Niño events.

Terra will make comprehensive measurements of phytoplankton biomass as well as dissolved organic matter in the upper ocean. These measurements of the "standing stock" of the base of the marine food web are critical to understanding the interactions between ocean circulation and productivity. Terra provides for the first time the ability to measure "fluorescence," or energy emitted by phytoplankton during photosynthesis. Fluorescence measurements will greatly improve the estimates of the rates of marine productivity.

The combination of both standing stock and growth rate will significantly improve our understanding of the ocean and climate. An Earth-orbiting NASA scatterometer will measure wind velocity over the ocean. These data will complement Terra by providing essential information on processes that drive ocean circulation. Winds affect the degree of ocean mixing as well as providing the momentum to drive ocean currents.

QuikSCAT will provide high-resolution maps of wind velocity to study specific processes such as coastal upwelling, as well as global-scale maps of winds to examine large-scale processes such as El Niño. Terra, and QuikSCAT provide oceanographers with unprecedented data for scientific studies of the air-sea couplings and their effects on the marine biosphere. With these data, scientists can refine their models of the ocean's physical, biological, and chemical influences on global climatic and environmental change.

5

Ice Ages and Global Warming

Increases in average world-wide temperature of the magnitude currently predicted can be cancelled by preventing about 1% of incoming solar radiation—insolation—from reaching the Earth. This could be done by scattering into space from the vicinity of the Earth an appropriately small fraction of total insolation.

Scattering Fundamentals

In general, three basic types of scatterers exist, for scattering any type of electromagnetic radiation, including sunlight. The simplest type is based on any material in which the electric fields of light cause a displacement of electric charges; thus, any material at all can be used. The magnitude of the displacement of charges by an electric field of unit strength is measured by the dielectric constant e, where $\varepsilon=1$ means there is no displacement. The scattering is proportional to $(\varepsilon-1)^2$, that is, highly polarisable materials generally will be more useful. This class of scatterer requires the near-optimal deployment of an estimated several million tons of scattering material in order to prevent an estimated (global- and time-) average temperature increase of 3±1.5°C associated with a doubling of atmospheric CO_2 during the coming century; the corresponding cost is ~$0.5 billion/year.

More effective scatterers can be realised by employing that subset of materials which exhibit high electrical conductivity. In this special case, electrons may be separated from their original locations by any distance, and it is the magnitude of the optical-frequency current carried by these electrons that characterise the effectiveness of such scattering materials—which are generally metals. Employed near-optimally, tens of thousands of tons of high-conductivity metal—roughly 1% of the required mass of dielectric materials—are required to scatter 1% of the Earth's total insolation; the corresponding costs are $0.07-0.14 billion/year. In principle, the most effective of all possible scatterers are atoms or molecules that scatter light in resonance. Such extremely strong scattering can be obtained for light of a frequency adapted to a specific atom or molecule.

The simplest example would be scattering of a narrow band of red light by lithium atoms or of yellow light by sodium atoms. Unfortunately, such exceptionally strong scatterings occur only in the immediate neighbourhood of an atomic transition-frequency, and the atom will selectively interact with light of frequencies which deviates from the resonant one by about one part in ten million. This difficulty can be overcome by broadening the resonance or by using scatterers that have many separate resonances—or, most effectively, by a combination of these two approaches. Of the order of 1 million tons of such resonant-type scattering material are estimated to suffice to remove 1% of the total insolation of the Earth; the corresponding cost may be $0.3-0.75 billion/year.

The intrinsic scattering strengths of dielectrics, electrical conductors, and resonant scatterers are in the approximate ratio of 1 to 10^4 to 10^6, respectively, for visible light; in practical implementations useful for insolation modulation, however, these ratios may be much different.

There are three obvious choices for deployment-sites for scatterers on scales of interest for insolation modulation. One is the terrestrial stratosphere, the second is in a low-Earth orbit, and the third is a position along the line between the centres of the Earth and the Sun.

Of the three deployments, the stratospheric location is by far the least expensive on a pound-for pound basis; positioning mass in the stratosphere currently is at least 10^4 times less costly than putting it into low Earth orbit. Moreover, the mid-stratospheric residence time of sub-microscopic scattering particles of anthropogenic and natural origins is comparable to the halfdecade residence time of its molecular components, so that appropriately fine-scale particulate loadings of the middle stratosphere will persist for five-year intervals.

However, the stratosphere is a chemically uncongenial location due to the high flux of ultraviolet radiation from the Sun and the presence of oxygen, particularly in the more reactive form of ozone. Deployment in low-Earth orbit is an obvious alternative, one which offers potentially very longterm positional stability combined with excellent durability of many materials. Technologies that could greatly decrease the cost of space-launch could make a telling difference in the practicality of all types of space-deployed scattering systems of scales appropriate to insolation modulation.

Light pressure arising from the momentum imparted to the scatterer by sunlight may significantly perturb the orbital elements of the scatterer, and managing this momentum poses an additional technical challenge to LEO-deployed scatterers. An interesting though not necessarily a practical case comprises the third alternative. Terrestrial insolation has an angular definition of one part in 120. Thus, if the scattering system is deployed ~10^2 times the Earth's diameter distant from the Earth, small-angle ($<1^{\circ}$) scattering will suffice for an appropriate deflection of the Earth-directed sunlight. This small angle permits the use of relatively very modest quantities of either conductors or dielectrics to comprise the scattering system -approximately 10^2-fold smaller than those needed to effect insolation modulation of the same magnitude when deployed near the Earth.

The management of the radial and angular momenta of the sunlight scattered poses basic, albeit quite tractable, issues with respect to position maintenance. There are obviously numerous ways in which the potentialities and difficulties mentioned above can give rise to a workable scattering system, one of a scale adequate to modulate the total insolation of the Earth by 1%.

Sub-Microscopic Oxide Particulates

During the present decade, the eruption of Mt. Pinatubo in the Philippines induced a transient drop in the global mean temperature of ~0.5° K, apparently due to insolation modulation by volcanic particulates. It is believed that this cooling was induced predominantly by scattering of sunlight by SO_2^- based particulates of sub-micron scale, ones which may have grown into more effective scatterers by scavenging residual stratospheric water and cations, resulting in myriad still-sub-micron droplets of high-concentration sulphur acids and salts. Indeed, it has been suggested that the advent of marked greenhouse effects due to CO_2 emissions has been delayed through the present time by the simultaneous emission of large quantities of sulphate particulates, resulting in significant tropospheric scattering of sunlight.

To these extents, the case of dielectric scattering-based insolation modulation already has some empirical basis.

It may well be feasible to transport and disperse enough SO_2 into the stratosphere to produce the desired insolation modulation effect—and even to do so partly on the basis of existing experience, as well as much prior analysis. It has also been suggested that alumina injected into the stratosphere by the exhaust of solid-rocket motors might scatter non-negligible amounts of sunlight. Introduction of scattering-optimised alumina particles into the stratosphere may well be overall competitive with use of sulphur oxides; alumina particles offer a distinctly different environmental impact profile.

Conducting Sheets

The reference 1% reduction in insolation might be obtained by deploying electrically-conducting sheeting, either in the stratosphere or in low Earth orbit. Following are three quite different physical mechanisms.

In the first, mixtures of suitable metals are deposited in ultra-thin layers and convenient area, and are then protectively coated. Platelets of such material are then deployed in the stratosphere—or perhaps in low Earth orbit—and act to absorb sunlight by the photoelectric effect; the absorbed energy is then thermally re-radiated, with ~half escaping into space.

In the second, metallic "nets" of ultra-fine mesh-spacing are employed to reflect solar photons of optical wavelengths into space.

In the third, optimised metallic-walled balloons—similar in concept to ones with which children play—are self-deployed into the stratosphere from ground level, where they serve to scatter insolation.

Each of these approaches involves total masses of the order of 10^5 tons, although the detailed mass budgets are quite different. The total cost-to-own any one of these three metallic systems at full-scale appears to lie in the range of \$0.07-0.14 billion/year.

Quasi-Resonant Scatterers

A third approach involves the use of resonant—actually quasi-resonant—scatterers, also deployed in the stratosphere. While the potential mass efficiency of this class of scatterers is singularly high, practical considerations centered on photochemical durability in the stratosphere indicate that total masses of 10^6 tons of material may have to be deployed—still modest-scale systems which moreover may have to be replaced only twice per decade.

Fine-Grained Insolation Modulation

Consistent with the slow latitudinal mixing-time of the stratosphere well above the tropopause, different amounts of scattering material might be deployed (e.g., at middle stratospheric altitudes, ~25 km) at different latitudes, so as to vary the magnitude of insolation modulation for relatively narrow latitudinal bands around the Earth, e.g., to reduce heating of the tropics by preferential loading of the mid-stratospheric tropical reservoir with insolation scatterer.

Indeed, scatterers of sunlight could be deployed at some latitudes to decrement insolation, while scatterers of Earth-emitted long-wavelength infrared radiation could be deployed at other latitudes. Differential cooling and heating, respectively, of underlying land-and-ocean latitudinal bands could thereby be accomplished. Furthermore, use of scatterers of varying stratospheric residence times to simultaneously modulate insolation and LWIR radiative losses in a specified latitude band might be employed to fine-tune, e.g., diurnal or seasonal temperature variability.

6

Impacts of Climate Change on Biodiversity

The climate of the Earth has never been stable, least of all during the history and evolution of life on Earth. Recent glacial periods, for example, have been (globally) 4°-5°C cooler than now, and some interglacials have been (perhaps) 1°-2°C warmer. These prehistoric changes in climate were clearly natural in origin and occurred on a planet inhabited by primitive societies with far smaller populations than at present. Indeed, the regularity of the diurnal and seasonal rhythms of our planet has always been overlain by inter-annual, multi-decadal and millennial variations in climate, over whatever timescale climate is defined.

Ecosystems and species have moved, often freely, in response to such past changes and have evolved within this climatic history. The causes of contemporary and future changes in climate, their rate and their potential significance for ecosystems and for the human species, however, are all notably different from anything that has occurred previously in history or pre-history. Evidence of global warming is plentiful and includes data on satellite-measured sea level change (2 mm rise per year from 1993-2003), lengthened growing season (by two to three weeks in the past 15-20 years) and increased precipitation intensities resulting in increased flood risk during winter in the northern hemisphere.

The causes are now dominated by human perturbation of the atmosphere and the rate of warming already exceeds anything experienced in the last 10,000 years. It is set to become more rapid than anything experienced in recorded human history and, given the ecological imprint made by our current and growing population of six billion and more, the significance of this prospect for the natural world and for human society is qualitatively different from previously experienced changes in climate.

The atmosphere delivers both resources (eg rain, sun, wind) and hazards (eg hurricanes, blizzards, droughts) to ecosystems and societies. Through autonomous and/or planned adaptation, species and individuals, ecosystems and societies, are fashioned to a considerable extent by these climatic constraints. Ecosystems and our human cultures and economies are 'tuned' to the climate in which they evolve. Yet highly successful cultures develop in varying climates: cold/dry climates (eg Finland), cold/wet climates (eg Iceland), hot/dry climates (eg Saudi Arabia) and hot/wet climates (eg Hong Kong).

All societies have therefore evolved strategies to cope with some intrinsic level of climatic variability—for example, nomadic pastoralism, flood prevention, building design, weather forecasting, early warning systems and the weather-hedging industry are all forms of human response to the variability of climate or the extremes of weather. Similarly, ecosystems are resilient to some extent of climatic variability, indeed may actually require it. Consequently, there exists some level of variation in climate or some frequency or severity of weather extremes that can be 'accommodated' using existing strategies or behaviour.

Exactly what can be accommodated, however, varies greatly within and between societies and ecosystems, so that vulnerability to weather and climate change is strongly differentiated. For example, in developed, northern nations, it is the elderly who die during heat waves, and in developing countries, it is often the poor and marginal who have their homes washed away in shanty towns built on flood plains.

So the central concern is not that humans are altering climate—we have modified our environment to a marked extent throughout our history—but whether these changes in climate can be accommodated using our existing capacity to adapt, drawing upon our intellectual, regulatory, social or financial capital and whether ecosystem resilience is large enough to survive these climatic perturbations given the other pressures they are subjected to by human development.

We face certainly continuing, probably accelerating and possibly unprecedented changes in the Earth's climate over the coming years and decades. These changes in such a fundamental resource for society, and in such a powerful influence on ecological, economic and cultural development, will introduce new challenges for the way we live with and influence climate. Some of these challenges may be broadly foreseeable, many of them may not. Some of the risks associated with a rapidly changing climate may be quantifiable, many of them may not. As evidence is emerging that some physical and biological systems are already reacting to this human induced change in climate, and as we know that, at least for some regions and for some communities and ecosystems, climate variability already imposes huge costs, doing nothing is unlikely to be the best option.

Societies need to develop and implement appropriate strategies to reduce the risks associated with a changing climate—to ensure that these changing climatic resources are appropriately exploited and that the adverse impacts of changing climatic hazards are minimised. Mitigation measures are required to reduce global greenhouse gas emissions with the intention of eventually stabilising atmospheric concentrations at some level at which an acceptable dynamic equilibrium could be sustained between climate, ecosystems and human society. What this level ought to be, however, remains poorly known and increasingly contested.

On the other hand, due to the inertia of both the climate system and our energy structures, greenhouse gases accumulated and accumulating in the atmosphere since the pre-industrial era will continue to affect global climate long into the future. Together with the existing exposure of many communities and assets to extremes of weather, adaptive measures become essential in order to enhance the coping abilities of valued ecosystems, vulnerable communities and exposed infrastructures.

These crucial perspectives about climate change need to be integrated fundamentally into the full range of policy measures that are demanded by our drive towards sustainable development, an argument equally valid for the nations of the south as for the nations of the north. We all need to come to terms with climate change.

Models of the Global Impact of Climate Change on Biodiversity

Change in biome area simulated by coupled Global Circulation Models (GCMs) and Global Vegetation Models (GVMs) may be used to estimate changes in species richness, if the assumption is made that species are faithful to their biome. That is, the biome's climatic definition is assumed to be a proxy for the climatic envelope of multiple resident species. Changes in biome areas were examined under two scenarios:

1. shifts in biomes kept pace with shifts in climatic conditions and
2. biomes fail to shift to new areas due to migration limitation.

In addition, estimated future migration rates are compared with post-glacial rates, and are used to locate areas that may be disproportionately important in facilitating future migration. Biome mapping was undertaken using 14 combinations of seven GCMs and two GVMs. Having determined the change in the area occupied by a biome, a species-area relationship was used to estimate the number of species present in a patch of a given biome and a given size. If there is a decline in the total area with change, the relationship would predict a loss of species within that patch. In this case, a continental z-value (species-area exponent) of 0.15 was used.

Biome changes were simulated using a GVM driven by a GCM climate. All models showed declines in the area of potential distribution of tundra, tundra/taiga and arid lands, and increases in temperate mixed forest, tropical broadleaf forest and grassland. Under the first scenario with perfect migration, all models showed declines in the areas of tundra and tundra/taiga, respectively ranging between 41-67% and 33-89% of the total area depending on the particular combination of GCM and GVM. Corresponding estimates of species loss were between 8-15% and 6-28% of the biota using a conservative species-area exponent (0.15).

Evidence of net declines in arid lands was also obtained. By contrast, temperate mixed forest showed consistent increases (49% on average) and tropical broadleaf forest and grassland also tended to show increases in area. Under the second scenario with no migration, all biome types declined in area, especially those at high latitudes and altitudes such as tundra, tundra/taiga, boreal conifer forest and temperate evergreen forest (55, 85, 46, 52% loss, respectively).

Even tropical broadleaf forest showed an 8% loss in area, corresponding to the possible loss of tens of thousands of species. Although this second scenario is unrealistic in that it assumes zero migration, it does highlight the potential for impacts in a diverse array of ecosystems, and the potential importance of migration in mitigating these impacts.

In order for the biome to successfully move to its new geographical niche, species would be required to migrate at least from the nearest point of the existing biome to each part of the new biome. This assumes that species occupy the whole of their biome space, which is more likely for widespread species. On average, the models showed that species in 19% of global grid cells would require a migration rate of over 1 km/year (the post-glacial rate of movement) to reach new suitable areas. The largest spatial shifts are seen in the northern hemisphere. At 1 km/year, Picea (spruce) would require 1,000 years rather than 100 years of migration to keep up with the shifting boreal biome.

Hotspots, under Myers' definition, occupy <2% of land area but hold 44% of plant and 35% of vertebrate species. When hotspot areas alone are considered, the range of response to the different scenarios is emphasised. Even when broader biome definitions are adopted, so increasing the modelled habitat breadth of individual species, potential for species loss is high. With narrow habitat breadth and low migration, the rates of species loss in the model approach those seen with tropical deforestation.

Tree/grass Dynamics

In southern Africa, fire-maintained systems have high biodiversity compared to the forest systems that appear with fire suppression. These C4 tropical grasslands and savannas appeared abruptly, between six and seven million years ago. Fire dynamics may have enabled their displacement of the early Tertiary's tall forests. Fire can be considered a generalist herbivore, which grasslands are able to regenerate from quickly. Young trees do not survive fire well, so savanna trees have fire-tolerant strategies. The sapling accumulates energy reserves at low height, and then shoots up rapidly to place its growing point above fire level. Saplings can be restricted to the understorey for decades.

The Sheffield dynamic global vegetation model includes a basic fire model, which performs well when compared with long-term burning experiments. When fire is switched off in the DGVM (Dynamic Global Vegetation Model), more forests are seen worldwide, in particular in Africa. After a fire, high levels of light, water and nutrients are available. CO_2 is then a limiting factor for woody plant growth. With increasing atmospheric CO_2 levels, woody plants reach fireproof height more rapidly. Changes in CO_2 concentration may therefore influence the savanna system, even before potential climate change impacts are considered.

Atmospheric CO_2 has rarely been higher than 280 ppm in the last 500,000 years, but post-industrial levels are 370 ppm and rising. A demographic model of tree response to burning was coupled to the DGVM. Tree numbers increased with CO_2 concentration. They were eliminated from savanna regions in the last glacial period, at low CO_2 concentrations. Some existed in the pre-industrial period, but more occurred under present-day conditions. These results match palynological records from the Wonderkrater savanna region—little tree pollen existed pre-Holocene here.

Savanna appears to be an interglacial phenomenon in South Africa. Savanna tree trials in CO_2 chambers at the National Botanical Institute also match the simulations, although growth rates at 550 ppm are not as high as seen in the physiological model. Long-term plots were set up in the Kruger National Park in 1954. More saplings have reached maturity in the last 20 years than previously. Worldwide, an increasing tree density is being seen in savannas, and CO_2 concentrations are 20% higher than in the 1950s.

In South Africa, savanna ecosystems host the most species richness and endemic species. A great change in grassland biota will be expected if this level of bush encroachment into fire-dependent systems continues. Fire policy here is contentious. Anthropogenic fires are still frequent, but are never started deliberately by managers. However, natural burns (caused by lightning) have been enabled to cross roads in order to reach fragmented parts of the savanna. However, a high-CO_2 world is not a 'natural world'. Two options are to manage for landscape heterogeneity, using patch mosaic burns, or to use frequent intense burns over broader areas to maintain mesic savanna vegetation in its current state. The effects of frequent burning on other savanna species such as invertebrates are not known.

Climate Change and Biodiversity in Tropical East Asian Forests

Tropical East Asia supports 20-25% of global biodiversity

and half a billion people. Until recently, the entire region was forested and plant and animal distributions largely reflected horizontal gradients in rainfall seasonality and a vertical gradient in temperature, both modified by edaphic and topographic factors, particularly towards their drier and colder ends. A latitudinal temperature gradient becomes increasingly important north of 18°N. Plant distributions probably also partly reflect dispersal history. The limited ability of current climate models to predict inter-annual (and inter-decadal) variability in rainfall patterns and the lack of information on current species-environment relationships makes it difficult to predict impacts of rapid climate change on biodiversity.

The most likely direct impact will be the loss of montane endemics because of exceeded temperature tolerances, reduced cloud water and the upward migration of pioneer species. In contrast, the direct impact of changes in rainfall patterns will probably be partly mitigated by adaptations to a rainfall regime that, even in 'normal' times, shows an exceptional degree of variability on all timescales. The direct impacts of climate change on biodiversity will probably be dwarfed by interactions with other human impacts. Tropical East Asia has the highest rates of deforestation, forest degradation and logging in the tropics, and the remaining forest is almost everywhere fragmented and disturbed.

Any increase in the frequency and/or intensity of dry periods would accelerate the existing synergy between logging, drought, agricultural clearance and devastating fires. In the longer term, outlying populations of wet forest species in dry areas, and vice versa, on edaphically or topographically extreme sites, will reduce the migration distances needed for equilibration, but fragmentation and the local extinction of dispersal agents will make migration impossible for most species. The evidence for migration deficits in current plant distributions suggests that migration rates for some species will be very slow even where continuous habitat is available.

The impacts of climate change on regional biodiversity can be mitigated only by the protection of large tracts of minimally disturbed forest that are continuous over both altitudinal and rainfall gradients. Existing protected areas are insufficient, so this will require both the rehabilitation of damaged forests and the reforestation of degraded areas. Planning at the regional scale needs to start now.

Climate Change and Amazonian Forest Biodiversity

Species' ranges are configured according to their tolerance of environmental conditions, especially climate, and their history of dispersal since speciation. Previous studies of the potential impact of climate change on biodiversity have been biased towards species of high latitudes. This situation results from a lack of detailed knowledge about the distribution of tropical biodiversity, and from the smaller degree of warming expected at low latitudes. However, various General Circulation Models (GCMs) simulate regional drying and increasing seasonality for parts of the tropics, including Amazonia. This may have a greater impact on tropical forest flora than temperature change alone.

The Amazon region holds a high proportion of global biodiversity, yet conservation plans rarely consider possible climate change impacts. An iterative selection routine was carried out to include species to represent the greatest practical range of Amazonian plant diversity, from the families Arecaceae, Balanophoraceae, Bignoniaceae, Caryocaraceae, Chrysobalanaceae, Fabaceae, Lauraceae, Proteaceae, Rubiaceae and Sapotaceae. Species were classified into plant functional types (PFTs), which share traits such as growth form and reproductive strategy. These species' ranges were projected from available spatial data and their response to climate change scenarios estimated.

Species' current distributions were modelled over a coarse scale (a 1° latitude-longitude grid), using a suitability index based on bioclimate variables. Distributions were additionally limited by species' absolute tolerances to extreme values, and by dispersal barriers. A size-structured population was simulated for each cell, to enable modelling of lags in response to climate change. In the standard impact scenario (SIS), future population processes were simulated over 100 years, with changes in the variables governing cell suitability being applied annually according to anomalies from a selected GCM.

In a reduced impact scenario (RIS), the run was repeated for each species using anomalies of half that magnitude. The range of potential outcomes for each species and PFT was evaluated. Widespread impacts were seen under both scenarios. An alternative impact scenario (AIS) was devised to examine the effects of allowing some competitive 'c-species' to thrive under heightened actual evapotranspiration (AET). The most vulnerable taxonomic groups, PFTs and geographical regions were identified as targets for monitoring and conservation action.

In particular, there is a dramatic loss of species' viability in much of northeastern Amazonia at 2095 under all scenarios. The far western part of Amazonia is identified as important for persistence of the greatest number of species. Areas falling between the major rivers of the region have very limited distribution data, so are highlighted for future biodiversity survey work.

Climate Change, Ocean Processes and Plankton Regime Shifts

The third report of the Intergovernmental Panel on Climate Change (IPCC) has shown that the rapid rise in mean global temperature seen in the last century was exceptional in the context of the last millennium. Mean surface (land and sea) global temperatures increased by 0.6°C ± 0.2°C in the 20th century and the heat content of the oceans, integrated to 3 km depth, has increased since 1950. There is now a scientific consensus that these increases are attributable to the even greater rate of increase seen in greenhouse gases. Surface temperatures are expected to increase by a further 1.4 to 5.8°C by 2100.

It is likely that changes in temperature at this scale and rate will have a pronounced effect on northern latitudes in the Arctic and, in turn, on the thermohaline circulation of the North Atlantic, which plays such a key role in the 'Global Conveyor Belt'. We may already be seeing evidence of these effects as the sea surface temperatures of the eastern Atlantic are highly correlated with northern hemisphere temperatures over the last 40 years. In the eastern Atlantic, evidence from the Continuous Plankton Recorder (CPR) survey indicates that plankton are able to integrate climatic signals that may not be the main forcing variable; as such, plankton may be used as an index of climate change.

Strong links have been demonstrated between the plankton and both Northern Hemisphere temperature (NHT) and the North Atlantic Oscillation (NAO), the dominant mode of atmospheric variability in the North Atlantic. The colour index of the CPR survey has shown a substantial increase in season length and intensity and implies increases in chlorophyll and primary production in a wide belt across the North Atlantic and especially in shelf seas. Parallel increases in the benthos imply that sedimentation from the plankton has also increased in the last decade.

At approximately the same time in the mid 1980s there appears to have been a step-wise change in physical, chemical and biological characteristics (at all trophic levels) in the North Sea that has been termed a regime shift. This is borne out by cluster analysis, which groups years according to both physical (sea surface temperature and NAO) and biological (phytoplankton, zooplankton and salmon catches) characters. This shows a distinct change from 1987 onwards. Changes in biogeographic zones and in the biodiversity of the plankton are linked to the decadal changes seen in plankton abundance with evidence for a northerly movement of warmer water plankton on the eastern side of the Atlantic and a southerly movement of plankton associated with colder water in the western Atlantic.

While climate variation may be considered to be natural, both the observed and predicted rates of climate change are unusually rapid and human effects on climate now appear to override natural variability. Ocean circulation and temperatures are changing in response. Physical and biological responses by oceans can be rapid and changes can be step-wise in the form of regime shifts. The northeast Atlantic may see large amplitude changes in the future.

Fisheries and Climate Change

Fisheries are the major cause of mortality for marine fish, once they have reached a catchable size. Most fish stocks have been declining for some time due to fishing. Nonetheless, climate fluctuations are also known to cause extensive shifts in species distribution and local biodiversity. Furthermore, climate change and fishing pressures may interact to exacerbate the risk of collapse of fish populations to below the level at which they can support fisheries. The demography of marine fish tends to differ from that of terrestrial vertebrates and freshwater fish in several important respects.

Marine fish have large population sizes with fewer boundaries to migration, high fecundity once mature, often have a dispersive plankton stage and they undergo large population fluctuations. There appear to be fewer barriers to migration in the sea. Hence, the constraints on distribution shifts in the sea are very different (and probably fewer) than on land or in freshwater systems, and marine populations probably maintain greater genetic exchange. Nevertheless, there is also concern about possible loss of genetic diversity (at least of genotypes), which may in turn reduce the capacity to adapt to climate change.

Fisheries' catch information provides much of our knowledge about fish populations. These data provide evidence for associations between climate change and marine fish abundance and distribution. The annual landings of cod in Greenland have varied enormously over the 20th century, ranging from very small local catches to almost 500,000 tonnes. This shows a close correlation with sea temperature fluctuations between 1 and 2.5°C, warmer temperatures being associated with bigger catches, but catches since the early 1970s have been lower than expected from temperature alone.

Changes in the abundance and distribution of other fish species in west Greenland are also associated with the period of warming from the 1920s. New species appeared from 1920 (Melanogrammus aeglefinus, Brosme brosme, Molva molva), rare species became common and extended their ranges (Pollachis virens, Salmo salar,

Squalus acanthias) and Arctic species no longer occurred in southern areas, but shifted their range northwards (Mallotius villosus, Gadus ogac, Reinhardtius hippoglossoides). Thus, climate can cause extensive shifts in species distribution and in local biodiversity.

In the North Sea, fishing has reduced life expectancy of several species, most notably plaice and sole, from around 10 years to two and cod, haddock and whiting from about four years to one year or less. Fishing tends to selectively remove large individuals and species. The Scottish August Groundfish Survey has shown that maximum fish size in the North Sea has declined by a factor of nearly eight between 1920 and 2000. The southern North Sea is at the edge of the cod's southern climatic range and the combination of warming and overfishing has led to a dramatic reduction in numbers.

Cod spawning biomass in the North Sea is at a historic low, but Barents Sea populations are still strong. Climate change and fisheries can impact on biodiversity is several ways. At the most obvious level, they can cause species extinction. They can cause stock extinction, which may or may not entail the loss of alleles or genotypes, and adverse changes in genotypic frequency. They can change distribution and abundance and we have most evidence for this. Climate change and sea level rise effects are probably severe for some sensitive systems, such as coral reefs and mangroves, and these will also affect fish populations. It is also becoming clear that the North Atlantic Oscillation has a strong, but not necessarily simple, effect on phytoplankton, zooplankton and fish dynamics.

Climate Change in Antarctic Marine Ecosystem

The Antarctic climate has changed throughout time and on a wide range of temporal scales. Southern Ocean bottom temperatures have changed dramatically over the past 60 Ma, as has been demonstrated by palaeotemperature studies of benthic foraminifera isolated from deep-sea sediment cores. Glaciation of the Antarctic continent probably began 36-40 Ma BP, and throughout its existence the continental ice-cap has varied in volume and the extent to which it has covered the continental shelves. A key driver of this variability has been Milankovitch cyclicity, long-term variations in the Earth's orbit which affect received solar radiation on timescales ranging from <20,000 to >100,000 years. These variations affect the timing and duration of seasons, and the extent of the tropical and polar regions.

For the last 15 Ma there has been consistent cooling of the Southern Ocean, and this is the period that has driven the evolution of the Antarctic climate and biota we observe today. Climatic and tectonic changes on geological and Milankovitch scales have produced a generally rich and diverse Antarctic marine fauna adapted to cope with this variability, by slow growth, longevity and intermittent recruitment. The marine fauna of Antarctica is generally rich and diverse, though some taxa are more speciose than others (notably species-rich groups include amphipod and isopod crustaceans, echinoderms and pycnogonids).

In contrast, the Southern Ocean has very few decapod crustaceans, and is dominated by only one group of teleost fish. Climate change is likely to influence this biodiversity through three processes, namely direct temperature effects on organisms, oceanographic shifts and changes in the dynamics of sea-ice. In some cases, any effects may be exacerbated by changes in UV flux as a result of changes in the seasonal development of the ozone hole. At present, there is evidence for small temperature changes in both shallow and deep waters around Antarctica, and for some associated oceanographic changes.

There is also some evidence for changes in sea-ice distribution and dynamics, though here it is difficult to distinguish a climate-change signal from natural long-period variability in the system. The clearest evidence for long-term changes in sea-ice dynamics comes from the Amundsen and Bellingshausen Seas, where there has been a decrease in winter sea-ice cover of about 10% since satellite records began. The atmospheric climate of the Antarctic Peninsula is warming faster than almost anywhere else on the planet, with warming being most evident in winter, and this may be related to the changes in sea-ice dynamics observed in the Amundsen and Bellingshausen Seas.

In contrast, the climate of the Antarctic continent shows no consistent pattern, and some places have even displayed a slight cooling. Data for biological responses to climate change in Antarctica are few, although there are strong indications of changes in population dynamics of some seabirds in relation to sea-ice. Climate change has already affected the breeding distribution of pygoscelid penguins, but no change has yet been observed for plankton or benthos.

Some marine invertebrates have been shown to live near their maximum temperature limits (a situation analogous to that of some tropical corals) and would therefore seem liable to extinction should seawater temperatures rise significantly in the near future. How the fate of individual sensitive species relates to the response of whole ecological assemblages is, however, far from clear.

The dynamics of sea-ice are complex, and can influence the population dynamics of many zooplankton which depend on sea-ice at stages in their life-cycle. An excellent example of this is the Antarctic krill, Euphausia superba, which is widely regarded as a keystone species in the food web at lower latitudes in the Southern Ocean. Young krill hatch along the western Antarctic Peninsula and are carried in the strong Antarctic Circumpolar Current to South Georgia, where they form a major part of the diet of many higher predators (fish, squid, seabirds and marine mammals). This means that variations in krill supply driven by changes in sea-ice dynamics thus influence the breeding success of dependent predators at South Georgia.

Demonstrating long-term changes in the biodiversity of marine systems is hindered by the paucity of data and the overwhelming effects of man's activities, even in Antarctica. Important insights into the potential effects of future climate change come from examining historical changes in the fauna, and from comparison of the two polar regions. To understand the impact of climate change on the Antarctic ecosystem as a whole, one needs to understand the structure of the food web. Although often portrayed as a simple, two-step linear food chain (diatoms to krill to whales, the classic Antarctic food chain of many textbooks), the Southern Ocean food web is similar to other marine food webs in that it is nonlinear, incorporates a microbial loop and has a significant flux to benthos.

The major difference from nonpolar food webs is the role of sea-ice. The food web structure and its non-linear dynamics make prediction of future responses very difficult. The removal of great whales might have resulted in a regime shift to a system dominated by fur seals and squid. Currently, we cannot predict the biological consequences of climate change on the Antarctic. Rather than focusing on individual taxa, the impact of climate change on assemblages needs to be investigated.

Climate Changes in East Antarctican Terrestrial Ecosystem

The global mean surface air temperature has increased by 0.19°C per decade from 1979 to 1998. Despite climate model predictions that show greenhouse gases will cause amplified warming in polar regions, clear signs of warming in Antarctica over the closing half of the 20th century have been confined to the Peninsula region. Ecological impacts on the Antarctic Peninsula have been attributed to this regional warming. In contrast, there is evidence of an annual cooling of 0.7°C per decade in the McMurdo Dry Valleys (77-78°S, 160-164°E) between 1986 and 2000. Fall and summer cooling of 2.0°C and 1.2°C per decade, respectively, dominate this trend. Analysis of Antarctic surface data from 1966 to 2000 suggests that cooling has been a dominant feature of Antarctic climate, outside of the Peninsula.

Cooling in the dry valleys was significantly related to decreased winds, and accompanied by increased clear-sky conditions. One consequence has been increased thickness of lake-ice. Ecosystem response to the cooling in the dry valleys was remarkably rapid and included decreased lake primary productivity (6-9%/year) and soil invertebrate numbers, especially soil nematodes (>10%/year). Despite these rapid responses to recent climate change, there is evidence that the characteristic biota of the dry valleys' ecosystem has persisted over very long timescales. There is evidence that past temperatures have been considerably cooler than at present so the system is likely to be resilient if current changes continue in the medium term.

One clear consequence of warming and cooling is a shift between a soil-dominant ecosystem and a lake-dominant ecosystem. During warm periods, lake levels rise and consume habitat for the soil organisms. Likewise, during cold periods, lakes dry up and soil habitat area increases. The actual response of the ecosystem to these changes has been the focus of research for the McMurdo Dry Valley Long Term Ecological Research (LTER) group since 1993. In addition to ongoing monitoring and experimental programmes, the LTER recently (November 2002) extracted long (up to 10 metres) sediment cores from the dry valley lakes so that direct evidence of how the lake ecosystems respond to changing climate will be available, allowing us to better predict response to future climate change.

Impacts of Climate Change on Arctic Vegetation

As part of global climate change, temperatures are predicted to increase by 2-6°C in Arctic latitudes by the end of the century, accompanied by an increase in precipitation, both changes being greatest in the winter months. Investigations of the history of vegetation in the Arctic demonstrate that when, during the Tertiary, the Earth was warmer than in the recent past, trees grew successfully at very high latitudes. Even during the past 10,000 years, the Arctic tree line was displaced substantially northwards during the early Holocene relative to its present position; temperatures at that time are estimated to have been c. 2°C warmer than in the recent past. At this time, the global extent of tundra was probably about 20% less than at present.

About 5,000 years ago, birch and pine woodland extended to the northern coast of Finnmark. The tree line retreated south about 2,000 years ago. The position of the Arctic tree line is of key importance, not only to other elements of the Arctic vegetation, but also to other organisms that are associated with the tundra biome that lies poleward of the tree line. Considerable changes to the abundance and distribution of tundra-adapted organisms would be expected if, as expected, global climate change leads to a reduction of the global extent of tundra by 50%. In addition, shifts of the tree line generate strong positive feedbacks to the climate system, especially via negative effects of tree cover on albedo that causes further warming and extension of suitable climate conditions for trees.

Field experiments in the Arctic indicate that the tree line may be inherently rather stable and resilient to relatively small fluctuations in climate, principally as a result of stabilising feedbacks at landscape scales. This leads us to expect a non-linear response when climatic change exceeds an as yet unpredictable threshold beyond which these stabilising feedbacks break down and positive feedbacks take over. A field experiment in which dead birch trees were erected on tundra as an artificial forest showed that this led to earlier bud burst of planted birch saplings within the forest and earlier snow melt.

Propagules of trees are already dispersed widely beyond the tree line, so migration rates are unlikely to much delay the spread of forests. Browsing by mammalian herbivores is also unlikely to have much impact, at least for birch, though pine and larch seedlings are selectively eaten by lemmings and voles. The change in feedback mechanisms will lead to a rapid shift to a new stable state in which the extent of tundra is markedly reduced and resultant positive feedbacks will lead to strengthened polar warming. Although the threshold for this effect is unknown, it is considered likely to be within about 2°C warming compared with the present.

Climate Change in the Tropical Montane Rainforests of Northern Australia

The endemic fauna of this area is particularly vulnerable to climate change due to a) its biogeographic history, having been restricted and therefore adapted to cool, wet and relatively a seasonal upland environments during past glaciations and b) the altitudinal gradient, ie species are adapted to a particular altitudinal niche and impacts of increasing temperatures are predicted to be most severe across altitudinal gradients. Altitude is the most important macroecological gradient, affecting almost every process. Bird species richness peaks at mid altitudes (c. 600-800 m). The effects of increases in temperature of between 1°C and 7°C on species distribution were assessed using bioclimatic modelling, based on a database of over 220,000 records from field samples and point locations.

Vertebrate distributions were modelled in relation to measures of temperature, rainfall and radiation using the bioclimate model BIOCLIM. Estimates were then made of the change in core range of each species under different climate scenarios, assuming that species continue to occupy the climate space they currently use. Models for 62 species of montane (>600 m altitude) endemics indicate that 1°C warming will result in an average 40% loss of potential core range, 3.5°C warming a 90% loss and 5°C warming a 97% loss. Warming of 7°C results in the loss of all potential core range for all species. About half of the montane endemics would lose all of their core potential range with 3.5°C warming.

Genetic diversity and evolutionary potential might also be lost, because there is considerable genetic diversity linked to the distribution of glacial refugia. Some of these refugia would cease to have suitable climate with even moderate warming. Insect and bird species richness and abundance are lower in areas with the greatest seasonality. Expected changes in seasonality include increased length of dry seasons, reduced climatic predictability and decreased moisture input. These are likely to affect insect biomass and, therefore, resources for insectivorous birds.

The physiology of many animals is not adapted to warmer temperatures. Some arboreal possums, highly adapted and poor dispersers, are prone to die of overheating and the eggs of microhylid frogs desiccate. Climate change is likely to cause catastrophic extinctions in the montane wet tropics due to range restrictions, increased fragmentation and physiological stress, resulting in large changes to community structure.

Climate Change and Habitat Fragmentation

In post-glacial times, it is known that some species changed their distributions rapidly in response to climatic amelioration, but they did not have to do so across landscapes dominated by human activities. Species show variable levels of success in shifting their ranges through fragmented landscapes in response to climate change. Whilst mobile generalists may increase their ranges, species with low dispersal rates, including sedentary habitat specialists, may not be able to track climate change. Recent studies of British butterfly species illustrate these trends.

Within a group of species which should all be increasing their ranges in Britain if they were fully occupying areas with suitable climate, mobile generalists such as the speckled wood and comma have increased their ranges by 24% and 30% respectively in recent years, whilst the silver-studded blue (a sedentary specialist) has declined by 28%. Habitat fragmentation reduces dispersal ability and contributes to lags in distributional change. The rate and direction of species dispersal is based on habitat availability (faster through more suitable landscapes), population density and growth, dispersal distances and the location of 'refuges'.

Habitat availability changes with climate and the habitat requirements of species change with climate;

indeed, some species show changes in habitat requirements to allow dispersal through patchy landscapes. For example, the brown argus has rapidly expanded its range since the early 1990s. It is no longer restricted to calcareous grasslands, where the common rock rose is its host plant, but is now using Geraniaceae as its host in many newly-colonised areas.

Distributional expansions are closely linked to habitat availability, and contractions exacerbated by fragmentation. The consequence is that biological communities will become increasingly dominated by certain types of animals and plants; mobile generalists may continue to prosper, whereas specialists are likely to continue to decline under the combined onslaught of habitat loss and climate change.

Impacts of Climate Change on Managed Landscapes

The Intergovernmental Panel on Climate Change (IPCC, 2001) concluded that 'the collective evidence indicates that there is a high confidence that recent regional changes in temperature have already had discernible impacts on many physical and biological systems'. Climate change is an important driving mechanism for altering the distribution of species and, therefore, the species composition of the habitats and ecosystems that characterise landscapes. The majority of European landscapes, however, are managed, often intensively, for a range of socio-economic purposes and are subject to both natural and anthropogenic pressures.

The ACCELERATES project (Assessing Climate Change Effects on Land use and Ecosystems: from Regional Analysis to The European Scale) is funded by the European Commission and is assessing the vulnerability of European agroecosystems to environmental change. This includes modelling the impacts of climate change on agriculture and biodiversity, both across the European Union and in selected habitats in six regions: the Ardennes, Belgium; East Anglia, UK; Northern Jutland, Denmark; Belluno Valley, Italy; Lesvos, Greece and Almeria, Spain.

The MONARCH project (Modelling Natural Resource Responses to Climate Change) is investigating the impacts of climate change on key species and habitats in Britain and Ireland. It has already shown that climate is one of several drivers of change affecting the distribution of species. The models are now being downscaled to provide landscape-scale guidance for setting nature conservation objectives in the context of climate change, and are being tested in four contrasting case study areas in the UK (Hampshire, Central Highlands, Snowdonia and Fermanagh/Tyrone/Donegal) selected for their susceptibility to climate change and conservation importance.

Dispersal modelling can simulate distributional changes driven by the ecology of the species and land use and climatic changes. Climate is important in influencing the distribution of species at the macro (European) scale, whereas land cover appears to dominate at the meso (regional) scale. Nature conservation policy needs to adopt a more holistic, dynamic approach to accommodate the changes in the distribution of species and the composition of habitats across landscapes brought about by climate change and other drivers of environmental change.

Biological Invasions and Climate Change

Biological invasions and climate change are two of the greatest environmental challenges that we face today. Individually, each of these elements of global change is the subject of much research. However, studies of invasion biology have rarely considered climate change, and vice versa. Such research is important because biological invasions can have strong effects on the structure and function of ecosystems that are responding to a changing climate and on the services provided by those ecosystems; climate change is also altering the context in which potentially invasive species succeed or fail. There is a variety of reasons why climate change might be expected to increase the success of biological invaders.

For instance, a rapidly changing climate might favour species that can extend their ranges quickly or that can tolerate a wide range of climatic conditions. Both of these traits are shared by many invasive plant species. As species shift ranges and habitat compositions change in response to climate change, animals that are generalists may have greater competitive success than specialists. Invasive animal species tend to be generalists, which may increase their success and threaten some native species. Few studies have focused on the general mechanisms through which climate change could benefit invasive species.

However, a number of studies have examined responses of the biota in a specific area to year-to-year differences in environmental conditions, or to experimental manipulations. Results from these studies can help us search for general patterns, and give us some indication of the type of responses that we can expect in the future. For example, the range of invasive species is directly influenced by climate and less by biotic interactions, whilst abundance is indirectly influenced by climate and more by resource availability.

Although few studies have addressed this topic, it is clear that some aspects of climate change will exacerbate problems with some invasive species. How much climate change will benefit invasive species remains difficult to predict and will always be difficult to measure. Hopefully, results from future research will teach us how to minimise the benefits that invasive species might draw from a changing climate.

Impacts of Climate Change on Mangrove Ecosystems

Mangrove forests are tropical intertidal habitats and are extensively developed on accretionary shorelines. There are 34 species of mangrove trees, in addition to three hybrids, with the centre of diversity being in southern Papua New Guinea. There is a decline in diversity from west to east across the Pacific, reaching a limit at American Samoa, while Atlantic mangrove species diversity is relatively poor. Global distribution is controlled by the 20°C isotherm, with diversity, forest height and associated productivity declining with increasing latitude away from the equator.

Sea level rise poses a major threat to mangrove ecosystems through sediment erosion, inundation stress and increased salinity at landward zones. These problems will be exacerbated for mangrove stands that are subjected to 'coastal squeeze', ie where landward migration is restricted by topography or human developments. Increased air temperatures and atmospheric CO_2 concentrations are also likely to increase mangrove productivity, change phenological patterns, and expand the ranges of mangrove forests into higher latitudes.

Mangroves can provide important services for adjacent ecosystems, and also supply many useful products to human societies. For example, mangroves provide

1. nursery habitats for many species of fish and invertebrates that spend their adult lives on coral reefs,

2. sediment trapping to sustain offshore water quality for coral reefs,
3. protection for inland sites from storm surges and flooding,
4. building materials,
5. traditional medicines,
6. firewood and
7. food.

As human populations have expanded, the shortage of productive land in many developing countries has led to the clearance of large areas of mangrove for agriculture and aquaculture production. Demands for timber (for charcoal, building etc) and coastal development space have also been highly damaging. Mangrove forests in some areas have been reduced to mere relicts of their former ranges as a result of human exploitation. In addition to these pressures, mangroves are threatened by sea level rise, projected between 0.9 and 8.8 mm/year.

Although there are several factors important in determining patterns of mangrove advance or retreat, studies have shown that mangroves are closely controlled by sea level elevation at their seaward margin. Mangrove species display a distinct zonation from low to high water, based on controls including the frequency of inundation and salinity exposure. This is controlled by the elevation of the substrate surface relative to mean sea level. Hence mangrove substrates can keep up with sea level rise through vertical accretion. Some of this accumulation will be from organic matter production, but this can be augmented by external inputs of sediment from rivers.

Rates of accretion reveal that mangrove ecosystems are highly vulnerable to projected rates of sea level rise.

Mangroves of low relief islands in carbonate settings that lack rivers are probably the most sensitive to sea level rise, owing to their sediment-poor environments and hence poor rates of vertical accretion. Mangrove response to sea level rise has been investigated by reconstructing Holocene analogues in the Cayman Islands, Tonga and Bermuda. Radiocarbon dating of stratigraphy determined a peat accretion rate of 1 mm/year for all locations. Recession of mangrove forests and replacement by lagoon environments are shown to occur during more rapid sea level rise.

On Grand Cayman, 20 km^2 of mangroves receded between 4080 and 3230 years BP. In Tonga, a large mangrove swamp that had persisted from 7000 to 5500 years BP, retreated when rates of sea level rise exceeded 1.2 mm/year. In Bermuda, present rates of sea level rise exceed 1.2 mm/year, and contemporary recession of the seaward margin of mangroves has been demonstrated.

The extensive coastal mangrove swamps of southern New Guinea (Irian Jaya) are also retreating from rising sea level. This demonstrates that, while low island mangroves are likely to be the most sensitive to sea level rise, continental margin mangroves will also suffer disruption and retreat. Mangroves have the capacity for extensive establishment under conditions of stable sea level, but are highly prone to retreat under conditions of sea level change.

Impacts of Arctic Zones and Low-lying Coastal Areas

Arctic zones and low-lying coastal areas are amongst the regions that are most vulnerable to the impacts of climate change and associated sea level rise. Migrant bird species that move between arctic and coastal temperate regions on an annual basis may thus face dramatic changes to the habitats that they use at both ends of the migratory range.

This is particularly true for the many high arctic breeding species that are almost entirely restricted to intertidal habitats in temperate zones during winter.

Coastal management in temperate zones for migratory birds must therefore take account of both breeding and winter season processes, and any interactions between the two. Changes in climate and sea level can potentially influence bird populations through a suite of direct and indirect routes in both the breeding and wintering seasons. For example, northward movement of the tree line in the arctic may reduce the area of tundra available for migrant birds at the same time that sea level rise and changing precipitation and temperature patterns may alter the structure or quality of temperate wintering areas.

Whereas climate change and sea level rise may be the primary drivers of change in the arctic, in temperate zones policy responses to climate change are likely to have a more direct and immediate bearing on biodiversity. It is thus critical that policy decisions in the coastal zone are informed by species-level studies that address the complexity of the processes influencing population responses to climate change. Biodiversity conservation in temperate coastal zones is structured through a network of site designations, underpinned by national and international legislative frameworks.

Decisions relating to the management and long-term sustainability of these sites are key in maintaining networks for migratory species. Coastal management for migratory species requires multi-disciplinary, integrated approaches in which models of structural changes to coastlines and consequent impacts on habitat structure and distribution are linked to models of species' responses to climate change and sea level rise. The Regional Coastal Simulator, which is being developed at the Tyndall Centre, aims to integrate these processes to provide a tool for coastal managers to explore the consequences of sea level rise and associated policy decisions in East Anglia.

Coral Reefs and Climate Change

Global climate change poses a substantial risk to the biodiversity, ecosystem functioning and productivity of coral reefs worldwide, and thus threatens their socio-economic value to human societies. In addition to impacts from climatic change, coral reefs are also under pressure from human activities (eg pollution, harvesting and coastal development) and natural stressors (eg crown-of-thorns starfish and disease). Historically, these ecosystems have typically been managed through marine protected area systems that focus on the threat from readily identifiable and 'tangible' anthropogenic activities such as fishing and development projects.

Incorporating the concept of climate change into the management of marine ecosystems is a relatively recent development that has yet to become widespread. However, it is increasingly evident that it is important to view the potential impacts of climate change in context with the other influences acting upon coral reef ecosystems. In terms of climate change, increasing sea temperatures are a matter of major concern for coral reefs throughout the world. Coral bleaching is a stress response where the algae symbionts (zooxanthellae) are ejected from the coral host, depriving the coral of nutrition from the products of photosynthesis.

Death of the host may result in severe cases. Bleaching can be triggered by a variety of stressors, but temperature related events are the most widespread. Coral reefs have already suffered major mortalities in many parts of the world as a result of high-temperature events, and there are projected increases in the number and magnitude of anomalously warm episodes. Increasing global sea temperatures may allow further expansion of coral reefs into the sub-tropics, but such processes are likely to be too slow to compensate for the loss of coral reefs to increased bleaching impacts associated with rising temperatures. Also expected to increase are the intensity and breadth of the destructive impact of extreme cyclones and flood plumes.

On a positive note, cyclones can be beneficial to coral reefs by encouraging mixing of surface and deep water, thereby cooling temperatures for a few days, which could be critical during times of harmfully warm waters. Rising sea level may promote growth of some shallow reef communities currently limited by water depth, but some deep areas may suffer from reduced light intensity as water depth increases, particularly where vertical accretion rates are reduced in stressed reefs. Increased atmospheric CO_2 will marginally reduce the alkalinity of reef waters, causing an increase in the rate of chemical dissolution of existing reef limestone, and a decrease in the deposition rate and strength of new limestone deposits.

The net effect of climate change on coral reefs will depend on their ability to adapt, but these ecosystems are generally regarded as already living near or at their thermal tolerance limits. Reef systems are highly heterogeneous in relation to the various influences acting upon them, which include responses and vulnerabilities to temperature-related bleaching events. When considering potential impacts of global warming, a marine park manager will want to know which areas of reef are more likely to escape heat stress or are more resistant to the impacts of climate-related bleaching. Coral bleaching research at AIMS includes spatial risk assessment and forecasting of ecological responses under various IPCC (Intergovernmental Panel on Climate Change) scenarios.

Modelling spatial variation in the likelihood of reef areas being exposed to harmfully high temperatures combines regional sea temperature predictions, local patchiness from satellite-derived sea surface temperatures and data on bleaching thresholds. The Great Barrier Reef suffered a major bleaching event in 2002, and systematic ecological assessments of this event provided data on taxonomic patterns of coral survival, injury and death. Mapping the future risk in terms of ecological impact and recovery will be relevant to short-term interests of the tourism industry, and also to management actions seeking to sustain the long-term ecological structure and functioning of coral reefs.

7

Climate Change and Human Health

The world's climate system is fundamental to the life support. Today, humankind's activities are altering the world's climate. We are increasing the atmospheric concentration of energy-trapping gases, thereby amplifying the natural "greenhouse effect" that makes the Earth habitable. These greenhouse gases (GHGs) comprise, principally, carbon dioxide, plus other heat-trapping gases such as methane, nitrous oxide and various human-made halocarbons.

During the twentieth century, world average surface temperature increased by approximately 0.6°C, and approximately two-thirds of that warming has occurred since 1975. Climatologists forecast further warming, along with changes in precipitation and climatic variability, during the coming century and beyond. Their forecasts are based on increasingly sophisticated global climate models, applied to plausible future scenarios of global greenhouse gas emissions that take into account alternative trajectories for demographic, economic and technological changes and evolving patterns of governance. The global scale of climate change differs fundamentally from the many other familiar environmental concerns that refer to localised toxicological or microbiological hazards.

Indeed, climate change signifies that, today, we are altering Earth's biophysical and ecological systems at the planetary scale—as is also evidenced by stratospheric ozone depletion, accelerating biodiversity losses, stresses on terrestrial and marine food producing systems, depletion of fresh water supplies, and the global dissemination of persistent organic pollutants. Human societies have had long experience of naturally-occurring climatic vicissitudes. The ancient Egyptians, Mesopotamians, Mayans, and European populations were all affected by nature's great climatic cycles. More acutely, disasters and disease outbreaks have occurred often in response to the extremes of regional climatic cycles such as the El Niño Southern Oscillation (ENSO) cycle.

Change in world climate would influence the functioning of many ecosystems and their member species. Likewise, there would be impacts on human health. Some of these health impacts would be beneficial. For example, milder winters would reduce the seasonal winter-time peak in deaths that occurs in temperate countries, while in currently hot regions a further increase in temperatures might reduce the viability of disease-transmitting mosquito populations. Overall, however, scientists consider that most of the health impacts of climate change would be adverse. Climatic changes over recent decades have probably already affected some health outcomes.

Indeed, the World Health Organisation estimated, in its "World Health Report 2002", that climate change was estimated to be responsible in 2000 for approximately 2.4% of worldwide diarrhoea, and 6% of malaria in some middle-income countries. However, small changes, against a noisy background of ongoing changes in other causal factors, are hard to identify. Once spotted, causal attribution is strengthened if there are similar observations indifferent population settings. The first detectable changes inhuman health may well be alterations in the geographic range and seasonality of certain infectious diseases—including vector-borne infections such as malaria and dengue fever, and food-borne infections which peak in the warmer months.

Warmer average temperatures combined with increased climatic variability would alter the pattern of exposure to thermal extremes and resultant health impacts, in both summer and winter. By contrast, the public health consequences of the disturbance of natural and managed food-producing ecosystems, rising sea-levels and]population displacement for reasons of physical hazard, land loss, economic disruption and civil strife, may not become evident for up to several decades. Unprecedentedly, today, the world population is encountering unfamiliar human-induced changes in the lower and middle atmospheres and world-wide depletion of various other natural systems.

Studying the impact of weather events and climate variability on human health requires appropriate specification of the meteorological "exposure". Weather and climate can each be summarised over various spatial and temporal scales. The appropriate scale of analysis, and the choice of any lag period between exposure and effect, will depend on the anticipated nature of the relationship. Much of the research requires long-term data sets with information about weather/climate and health outcome on the same spatial and temporal scales. For example, it has proven difficult to assess how climate variability and change has influenced the recent spread of malaria in African highlands because the appropriate health, weather and other relevant data have not been collected in the same locations and on the same scales.

In all such research, there is a need to accommodate the several types of uncertainty that are inherent in these studies. Predictions about how complex systems such as regional climate systems and climate-dependent ecosystems will respond when pushed beyond critical limits are necessarily uncertain. Likewise, there are uncertainties about the future characteristics, behaviours and coping capacity of human populations.

International Consensus on the Science of Climate and Health

In the early 1990s there was little awareness of the health risks posed by global climate change. This reflected a general lack of understanding of how the disruption of biophysical and ecological systems might affect the longer-term wellbeing and health of populations. There was little awareness among natural scientists that changes in their particular objects of study—climatic conditions, biodiversity stocks, ecosystem productivity, and so on - were of potential importance to human health. Indeed, this was well reflected in the meagre reference to health risks in the first major report of the UN's Intergovernmental Panel on Climate Change (IPCC), published in 1991.

The IPCC was established by WMO and UNEP in 1988. The IPCC's role is to assess the world's published scientific literature on:

(i) how human-induced changes to the lower atmosphere, via the emission of greenhouse gases, have influenced and are likely to influence world climatic patterns;

(ii) how this does, and in future would, affect various systems and processes important to human societies; and

(iii) the range of economic and social response options available to policy-makers to avert climate change and to lessen its impacts.

The IPCC's work has been done by many hundreds of scientists, worldwide. On a five-yearly basis, national governments propose scientists with expertise in the many topic areas included within this comprehensive review task. Topic review teams are then chosen to ensure proper geographic and disciplinary representation. Excluding the small number of scientists working at IPCC secretariat level, all this work of reviewing, discussing and writing is contributed voluntarily. The IPCC's draft assessments are subject to a series of internal and external peer-review processes. The final wording of IPCC report summaries are subject, via formal international conferences, to detailed and systematic scrutiny by governments.

The IPCC's Assessment of Health Impacts

In its Third Assessment Report the IPCC concluded that: "Overall, climate change is projected to increase threats to human health, particularly in lower income populations, predominantly within tropical/subtropical countries." That summary went on to state: "Climate change can affect human health directly and indirectly through changes in the ranges of disease vectors (e.g., mosquitoes), water-borne pathogens, water quality, air quality, and food availability and quality. The actual health impacts will be strongly influenced by local environmental conditions and socio-economic circumstances, and by the range of social, institutional, technological, and behavioural adaptations taken to reduce the full range of threats to health."

Broadly, a change in climatic conditions can have three kinds of health impacts:

— Those that are relatively direct, usually caused by weather extremes. The health consequences of various processes of environmental change and ecological disruption that occur in response to climate change.

— The diverse health consequences

— traumatic, infectious, nutritional, psychological and other—that occur in demoralised and displaced populations in the wake of climate-induced economic dislocation, environmental decline, and conflict situations.

Our understanding of the impacts of climate change and variability on human health has increased considerably in recent years. However, several basic issues complicate this task:

— Climatic influences on health are often modulated by interactions with other ecological processes, social conditions, and adaptive policies. In seeking explanations, a balance must be sought between complexity and simplicity.

— There are many sources of scientific and contextual uncertainty. The IPCC has therefore sought to formalise the assessment of level of confidence attaching to each health impact statement.

— Climate change is one of several concurrent global environmental changes that simultaneously affect human health—often interactively. A good example is the transmission of vector-borne infectious diseases, which is jointly affected by climatic conditions, population movement, forest clearance and land-use patterns, biodiversity losses (e.g., natural predators of mosquitoes), freshwater surface configurations, and human population density.

The IPCC concluded, with high confidence, that climate change would cause increased heat-related mortality and morbidity, decreased cold-related mortality in temperate countries, greater frequency of infectious disease epidemics following floods and storms, and substantial health effects following population displacement from sea level rise and increased storm activity. For each potential impact of climate change, certain groups will be particularly vulnerable to disease and injury. The vulnerability of a population depends on factors such as population density, level of economic development, food availability, income level and distribution, local environmental conditions, pre-existing health status, and the quality and availability of public health care.

For instance, those most at risk of being harmed by thermal extremes include socially isolated city dwellers, the elderly and the poor. Populations living at the present margins of malaria and dengue, without effective primary healthcare, will be the most susceptible if these diseases expand their geographic range in a warmer world. The IPCC report also underscores that our understanding of the links between climate, climate change and human health has increased considerably over the last ten years. However, there are still many gaps in knowledge about likely future patterns of exposure to climatic environmental changes, and about the vulnerability and adaptability of physical, ecological and social systems to such climate change.

Issues in Predicting the Health Impacts of Climate Change

The challenges in identifying, quantifying and predicting the health impacts of climate change entail issues of scale, "exposure" specification, and the elaboration of often complex and indirect causal pathways.

First, the geographic scale of climate-related health impacts and the typically wide time spans are unfamiliar to most researchers. Epidemiologists usually study problems that are geographically localised, have relatively rapid onset, and directly affect health. The individual is usually the natural unit of observation.

Second, the "exposure" variable—comprising weather, climate variability and climate trends—poses difficulties. There is no obvious "unexposed" group to act as baseline for comparison. Indeed, because there is little difference in weather/climate exposures between individuals in the same geographic locale, comparing sets of persons with different "exposures" is usually precluded. Rather, whole communities or populations must be compared—and, in so doing, attention must be paid to intercommunity differences invulnerability. For example, the excess death rate during the severe 1995 Chicago heat wave varied greatly between neighbourhoods because of differences in factors such as housing quality and community cohesion.

Third, some health impacts occur via indirect and complex pathways. For example, the effects of temperature extremes on health are direct. In contrast, complex changes in ecosystem composition and functioning help mediate the impact of climatic change on transmission of vector-borne infectious diseases and on agricultural productivity.

A final challenge is the need to estimate health risks in relation to future climatic-environmental scenarios. Unlike most recognised environmental health hazards, much of the anticipated risk from global climate change lies years to decades into the future.

While much health-impacts research focuses on future risk, empirical studies referring to the recent past and present are important. Standard observational epidemiological methods can illuminate the health consequences of local climatic trends in past decades—if the relevant data-sets exist. Such information enhances our capacity subsequently to estimate future impacts. Meanwhile, we should also seek evidence of the early health effects of climate change, since change has been underway for several decades.

The health impacts of future climate change, including changes in climatic variability, can be estimated in two main ways. First, we can extrapolate from analogue studies that treat recent climatic variability as a foretaste of climate change. Second, we can use predictive computer models based on existing knowledge about relationships between climatic conditions and health outcomes. Such models cannot predict exactly what will happen, but they indicate what would occur if certain future climatic conditions were fulfilled. The five main tasks for researchers are:

Establishing baseline relationships between weather and health: There are many unresolved questions about the sensitivity of particular health outcomes to weather, climate variability, and climate-induced environmental changes. For example, the major pathogens that cause acute gastroenteritis multiply faster in warmer conditions. Do higher ambient temperatures cause more illness? Apparently so—as is evident from the monthly salmonella infection count in NewZealand in relation to average monthly temperature.

Seeking evidence of early effects of climate change: There have been many, coherent, observations on physical and ecological changes attributable to recent global warming—but few indications yet of human health effects. Amongst these are changing patterns of infectious disease. Health researchers must allow for the fact that humans have many coping strategies, ranging from planting shade trees, to changing work-hours, to installing air-conditioning.

The challenge is to pick the settings, populations and health outcomes with the best chance of: (i) detecting changes, and (ii) attributing some portion of these to climate change. Impacts are likely to be clearest where the exposure outcome gradient is steepest, the local population's adaptive capacity is weakest, and when there are few competing explanations for observed relationships.

Scenario-based predictive models: Unlike most other environmental exposures, we know that the world's climate will continue to change for at least several decades. Climatologists now can satisfactorily model the climatic consequences of future scenarios of greenhouse gase missions. By linking these climate scenarios with health impact models, we can estimate the likely impacts on health. Some health impacts are. We need models with sufficient representation of the multi-faceted future world to provide useful, or credible, estimates of future health risks. Where possible, we should use a high level of "integration" to achieve realistic modelled forecasts of impact in a world that will have undergone various other demographic, economic, technological and social changes.

Evaluating adaptation options : Adaptation means taking steps to reduce the potential adverse impact of environmental change.

Estimating the co-incidental benefits and costs of mitigation and adaptation: Steps to reduce GHG emissions (mitigation) or to lessen health impacts (adaptation) may have other coincidental health effects. For example, promotion of public transport relative to private vehicles may not only reduce CO_2 emissions, but also improve public health in the near-term by reducing airpollution and road traffic injuries and increasing physical activity. Information about these "ancillary" costs and benefits is important for policy-makers. Note, however, for impacts that are either deferred in time or that extend into the distant future, the costing is not straight forward.

Health Impacts of Extremes Climate

Extreme climate events are expected to become more frequent with climate change. These disruptive events have their greatest impact in poor countries. The two categories of climatic extremes are:

- Simple extremes of climatic statistical ranges, such as very low or very high temperatures
- Complex events: droughts, floods, or hurricanes

The Pacific-based El Niño-Southern Oscillation (ENSO), an approximately semi-decadal cycle, influences much of the world's regional weather patterns. Climate change is likely to increase the frequency and/or amplitude of ElNiño. It illustrates well how climatic extremes can affect human health.

Climate, Weather, El Niño and Infectious Diseases

Both temperature and surface water have important influences on the insect vectors of vector-borne infectious disease. Of particular importance are vector mosquito species, which spread malaria and viral diseases such as dengue and yellow fever. Mosquitoes need access to stagnant water in order to breed, and the adults need humid conditions for viability. Warmer temperatures enhance vector breeding and reduce the pathogen's maturation period within the vector organism. However, very hot and dry conditions can reduce mosquito survival.

Malaria, today, is mostly confined to tropical and subtropical regions. The disease's sensitivity to climate is illustrated by desert and highland fringe areas where higher temperatures and/or rainfall associated with El Niño may increase transmission of malaria. In areas of unstable malaria in developing countries, populations lack protective immunity and are prone to epidemics when weather conditions facilitate transmission.

Dengue is the most important arboviral disease of humans, occurring in tropical and subtropical regions, particularly in urban settings. ENSO affects dengue occurrence by causing changes in household water storage practices and in surface water pooling. Between 1970 and

1995, the annual number of dengue epidemics in the South Pacific was positively correlated with La Niña conditions.

Rodents, which proliferate intemperate regions following mild wet winters, act as reservoirs for various diseases. Certain rodent borne diseases are associated with flooding, including leptospirosis, tularaemia and viral haemorrhagic diseases. Other diseases associated with rodents and ticks, and which show associations with climatic variability, include Lyme disease, tick borne encephalitis, and hantavirus pulmonary syndrome.

Many diarrhoeal diseases vary seasonally, suggesting sensitivity to climate. In the tropics diarrhoeal diseases typically peak during the rainy season. Both floods and droughts increase .the risk of diarrhoeal diseases. Major causes of diarrhoea linked to heavy rainfall and contaminated water supplies are: cholera, cryptosporidium, E.coli *infection*, giardia, shigella, typhoid, and viruses such as hepatitis A.

Heatwaves and Cold Spells

Extremes of temperature can kill. In many temperate countries, death rates during the winter season are 10-25% higher than those in the summer. In July 1995, a heat wave in Chicago, US, caused 514 heat related deaths and 3300 excess emergency admissions. Most of the excess deaths during times of thermal extreme are in persons with preexisting disease, especially cardiovascular and respiratory disease. The very old, the very young and the frail are most susceptible. In terms of the amount of life lost, the mortality impact of an acute event such as a heat wave is uncertain because an unknown proportion of deaths are in susceptible persons who would have died in the very near future.

Global climate change will be accompanied by an increased frequency and intensity of heatwaves, as well as warmer summers and milder winters. Predictive modelling studies, using climate scenarios, have estimated future temperature-related mortality. For example, the annual excess summer-time mortality attributable to climate change, by 2050, is estimated to increase several-fold, to between 500-1000 for New York and 100-250 for Detroit, assuming population acclimatisation. Without acclimatisation the impacts would be higher.

The extent of winter-associated mortality directly attributable to stressful weather is less easy to determine. In temperate countries undergoing climate change, a reduction in winter deaths may out number the increase in summer deaths. Without better data, the net impact on annual mortality is difficult to estimate. Further, it will vary between populations.

Impacts of Weather Disasters

The effects of weather disasters on health are difficult to quantify, because secondary and delayed consequences are poorly reported. El Niño events influence the annual toll of persons affected by natural disasters. Globally, disasters triggered by droughts occur especially during the year after the onset of El Niño. Globally, natural disaster impacts have been increasing. An analysis by the reinsurance company Munich Re found a tripling in the number of natural catastrophes in the last ten years, compared to the 1960s. This reflects global trends in population vulnerability more than an increased frequency of extreme climatic events.

Developing countries are poorly equipped to deal with weather extremes, even as the population concentration increases in high-risk areas like coastal zones and cities. Hence, the number of people killed, injured or made homeless by natural disasters has been increasing rapidly. Table 1. shows the numbers of events, deaths and people affected by extreme climatic and weather events in the past two decades, by geographic region.

Table 1. Numbers of extremc weather events, people killed and affected, by region of the world, in the 1980s and 1990s

1980s	Events	Killed (thousands)	Affected (millions)
Africa	243	417	137.8
Eastern Europe	66	2	0.1
Eastern Mediterranean	94	162	17.8
Latin America and Caribbean	265	12	54.1
South East Asia	242	54	850.5
Western Pacific	375	36	273.1
Developed	563	10	2.8
Total	1,848	692	1,336

1990s	Events	Killed (thousands)	Affected (millions)
Africa	247	10	104.3
Eastern Europe	150	5	12.4
Eastern Mediterranean	139	14	36.1
Latin America and Caribbean	298	59	30.7
South East Asia	286	458	427.4
Western Pacific	381	48	1,199.8
Developed	577	6	40.8
Total	2,078	601	1,851

Climate Changes and Infectious Diseases

Humans have known that climatic conditions affect epidemic diseases from long before the role of infectious agents was discovered, late in the nineteenth century. Roman aristocrats retreated to hill resorts each summer to avoid malaria. South Asians learnt early that, in high summer, strongly curried foods were less likely to cause diarrhoea. Infectious agents vary greatly in size, type and mode of transmission.

There are viruses, bacteria, protozoa and multicellular parasites. Those microbes that cause "anthroponoses" have adapted, via evolution, to the human species as their primary, usually exclusive, host. In contrast, non-human species are the natural reservoir for those infectious agents that cause "zoonoses" (Fig 1).

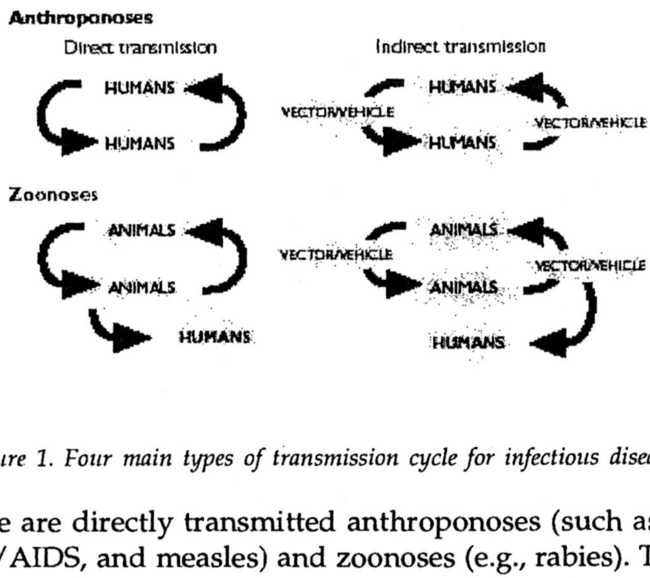

Figure 1. Four main types of transmission cycle for infectious diseases

There are directly transmitted anthroponoses (such as TB, HIV/AIDS, and measles) and zoonoses (e.g., rabies). There are also indirectly-transmitted, vector borne, anthroponoses (e.g., malaria, dengue fever, yellow fever) and zoonoses (e.g. bubonic plague and Lyme disease).

Vector-borne and water-borne diseases: Important determinants of vector borne disease transmission include:

1. vector survival and reproduction,
2. the vector's biting rate, and

3. the pathogen's incubation rate within the vector organism.

Vectors, pathogens and hosts each survive and reproduce within a range of optimal climatic conditions: temperature and precipitation are the most important, while sea level elevation, wind, and daylight duration are also important. Human exposure to water borne infections occurs by contact with contaminated drinking water, recreational water, or food. This may result from human actions, such as improper disposal of sewage wastes, or be due to weather events. Rainfall can influence the transport and dissemination of infectious agents, while temperature affects their growth and survival.

There are three categories of research into the linkages between climatic conditions and infectious disease transmission. The first examines evidence from the recent past of associations between climate variability and infectious disease occurrence. The second looks at early indicators of already-emerging infectious disease impacts of long term climate change. The third uses the above evidence to create predictive models to estimate the future burden of infectious disease under projected climate change scenarios.

Historical Evidence: There is much evidence of associations between climatic conditions and infectious diseases. Malaria is of great public health concern, and seems likely to be the vector-borne disease most sensitive to long-term climate change. Malaria varies seasonally in highly endemic areas. The link between malaria and extreme climatic events has long been studied in India, for example. Early last century, the river-irrigated Punjab region experienced periodic malaria epidemics. Excessive monsoon rainfall and high humidity was identified early on as a major influence, enhancing mosquito breeding and survival. Recent analyses have shown that the malaria epidemic risk increases around five-fold in the year after an El Niño event.

Early Impacts of Climate Change: These include several infectious diseases, health impacts of temperature extremes and impacts of extreme climatic and weather events.

Predictive Modeling: The main types of models used to forecast future climatic influences on infectious diseases include statistical, process-based, and landscape-based models. These three types of model address somewhat different questions. Statistical models require, first, the derivation of a statistical (empirical) relationship between the current geographic distribution of the disease and the current location specific climatic conditions. This describes the climatic influence on the actual distribution of the disease, given prevailing levels of human intervention.

By then applying this statistical equation to future climate scenarios, the actual distribution of the disease in future is estimated, assuming unchanged levels of human intervention with in any particular climatic zone. These models have been applied to climate change impacts on malaria, dengue fever and, within the USA, encephalitis. For malaria some models have shown net increases in malaria over the coming half century, and others little change.

Process-based (mathematical) models use equations that express the scientifically documented relationship between climatic variables and biological parameters e.g., vector breeding, survival, and biting rates, and parasite incubation rates. In their simplest form, such models express, via a set of equations, how a given configuration of climate variables would affect vector and parasite biology and, therefore, disease transmission. Such models address the question: "If climatic conditions alone change, how would this change the potential transmission of the disease?"

Using more complex "horizontal integration", the conditioning effects of human interventions and social contexts can also be incorporated. This modelling method has been used particularly for malaria and dengue fever. The malaria modelling shows that small temperature increases can greatly affect transmission potential. Globally, temperature increases of 2-3°C would increase the number of people who, in climatic terms, are at risk of malaria by around 3-5%, i.e. several hundred million. Further, the seasonal duration of malaria would increase in many currently endemic areas.

Since climate also acts by influencing habitats, landscape based modeling is also useful. This entails combining the climate-based models described above with the rapidly-developing use of spatial analytical methods, to study the effects of both climatic and other environmental factors (e.g. different vegetation types—often measured, in the model development stage, by ground-based or remote sensors). This type of modelling has been applied to estimate how future climate-induced changes in ground cover and surface water in Africa would affect mosquitoes and tsetse flies and, hence, malaria and African sleeping sickness.

How Much Disease would Climate Change Cause?

The global burden of disease attributable to climate change has recently been estimated as part of a comprehensive World Health Organisation project. This projects ought to quantify disease burdens attributable to 26 environmental, occupational, behavioural and lifestyle risk factors in 2000, and at selected future times up to 2030.

The disease burden comprises the total amount of disease or premature death within the population. To compare burden fractions attributable to several different risk factors requires, first, knowledge of the severity/disability and duration of the health deficit, and, second, the use of standard units of health deficit. The widely used Disability-Adjusted Life Year (DALY2) is the sum of:

— years of life lost due to premature death (YLL)

— years of life lived with disability (YLD).

YLL takes into account the age at death. YLD takes into account disease duration, age at onset, and a disability weight reflecting the severity of disease. To compare the attributable burdens for disparate risk factors we need to know:

1. the baseline burden of disease, absent the particular risk factor,
2. the estimated increase in risk of disease/death per unit increase in risk factor exposure (the "relative risk"), and
3. the current or estimated future population distribution of exposure.

The avoidable burden is estimated by comparing projected burdens under alternative exposure scenarios. Disease burdens have been estimated for five geographical regions. The attributable disease burden has been estimated for the year 2000. For the years 2010, 2020 and 2030, the climate-related relative risks of each health outcome under each climate change scenario, relative to the situation if climate change did not occur, were estimated. The baseline scenario is 1990. The future exposure scenarios assume the following projected GHG emission levels:

1. Unmitigated emission trends
2. Emissions reduction, achieving stabilisation at 750 ppm CO_2^- equivalent by 2210

3. More rapid emissions reduction, stabilising at 550 ppm CO_2^- equivalent by 2170

Some of the health outcomes associated with climate change are addressed in Table 2. These were selected on the basis of: (a) sensitivity to climate variation, (b) predicted future importance, and (c) availability/feasibility of quantitative global models. Additional likely health impacts that are currently not quantifiable include those due to:

— changes in air pollution and aeroallergen levels

— altered transmission of other infectious diseases

— effects on food production via climatic influences on plant pests and diseases

— drought and famine

— population displacement due to natural disasters, crop failure, water shortages

— destruction of health infrastructure in natural disasters

— conflict over natural resources

— direct impacts of heat and cold (morbidity).

Table 2. Health outcomes associated with climate change

Type of outcome	*Outcome*	*Incidence/Prevalence*
Food and water-borne disease	Diarrhoea episodes	Incidence
Vector-borne disease	Malaria cases	Incidence
Natural disasters	Fatal unintentional injuries	Incidence
Risk of malnutrition	Non-availability of recommended daily calorie intake	Prevalence

All independently-published models linking climate change to quantitative, global, estimates of health impacts were reviewed. Where global models do not exist, local or regional projections were extrapolated. Models were selected according to their assessed validity. Linear interpolation was used to estimate relative risks for inter-scenario years. Climate change will affect the pattern of deaths from exposure to high or low temperatures. However, the effect on actual disease burden cannot be quantified, as we do not know to what extent deaths during thermal extremes are in sick/frail persons who would have died soon anyway.

In 2030 the estimated risk of diarrhoea will be up to 10% higher in some regions than if no climate change occurred. Since few studies have characterised this particular exposure-response relationship, these estimates are uncertain. Estimated effects on malnutrition vary markedly among regions. By 2030, the relative risks for unmitigated emissions, relative to no climate change, vary from a significant increase in the South-East Asia region to a small decrease in the Western Pacific.

Overall, although the estimates of changes in risk are somewhat unstable because of regional variation in rainfall, they refer to a major existing disease burden entailing large numbers of people. The estimated proportional changes in the numbers of people killed or injured in coastal floods are large, although they refer to low absolute burdens. Impacts of inland floods are predicted to increase by a similar proportion, and would generally cause a greater acute rise in disease burden. While these proportional increases are similar in developed and developing regions, the baseline rates are much higher in developing countries.

Changes in various vector-borne infectious diseases are predicted. This is particularly so for malaria in regions bordering current endemic zones. Smaller changes would occur in currently endemic areas. Most temperate regions would remain unsuitable for transmission, because either they remain climatically unsuitable (e.g., most of Europe) or socioeconomic conditions are likely to remain unsuitable for reinvasion (e.g., southern United States). Uncertainties relate to how reliable is extrapolation between regions, and to whether potential transmission will become actual transmission.

Application of these models to current disease burdens suggests that, if our understanding of broad relationships between climate and disease is realistic, then climate change may already be affecting human health. The total current estimated burden is small relative to other major risk factors measured under the same framework. However, in contrast to many other risk factors, climate change and its associated risks are increasing rather than decreasing over time.

Ozone Depletion and Health

Scientists 100 years ago would have been incredulous at the idea that, by the late twentieth century, humankind would be affecting the stratosphere. Yet, remarkably, human-induced depletion of stratospheric ozone has recently begun—after 8,000 generations of Homosapiens. Stratospheric ozone absorbs much of the incoming solar ultraviolet radiation (UVR), especially the biologically more damaging, shorter-wavelength, UVR. We now know that various industrial halogenated chemicals such as the chlorofluorocarbons and methyl bromide, while inert at ambient Earth-surface temperatures, react with ozone in the extremely cold polar stratosphere. This destruction of ozone occurs especially in late winter and early spring.

During the 1980s and 1990s at northern mid-latitudes, the average year-round ozone concentration declined by around 4% per decade: over the southern regions of Australia, NewZealand, Argentina and South Africa, the figure approximated 6-7%. Estimating the resultant changes in actual ground-level ultraviolet radiation remains technically complex. However, exposures at northern mid latitudes, for example, are likely to peak around 2020, with an estimated 10% increase in effective ultraviolet radiation relative to 1980s levels.

In the mid-1980s, governments recognised the emerging hazard from ozone depletion. The Montreal Protocol of 1987 was adopted, widely ratified, and the phasing out of major ozone destroying gases began. The protocol was tightened in the 1960s. Scientists anticipate a slow but near-complete recovery of stratospheric ozone by the middle of the twenty-first century.

Types of Health Impacts of Ozone Depletion

The range of certain or possible health impacts of stratospheric ozone depletion are listed in Table 3.

Table 3. Possible effects of solar ultraviolet radiation on human health

Effects on skin

— Malignant melanoma

— Non-melanocytic skin cancer –basal cell carcinoma, squamous cell carcinoma

— Sunburn

— Chronic sun damage

— Photodermatoses

Effects on the eye

— Acute photokeratitis and photoconjunctivitis

— Climatic droplet keratopathy

— Pterygium

— Cancer of the cornea and conjunctiva

— Lens opacity (cataract) – cortical, posterior subcapsular

— Uveal melanoma
— Acute solar retinopathy
— Macular degeneration

Effect on immunity and infection

— Suppression of cell mediated immunity
— Increased susceptibility to infection
— Impairment of prophylactic immunisation
— Activation of latent virus infection

Other effects

— Cutaneous vitamin D production
    — prevention of rickets, osteomalacia and osteoporosis
    — possible benefit for hypertension, ischaemic heart disease and tuberculosis
    — possible decreased risk for schizophrenia, breast cancer, prostate cancer
    — possible prevention of Type 1 diabetes
— Altered general well-being
    — sleep/wake cycles
    — seasonal affective disorder
    — mood

Indirect effects

— Effects on climate, food supply, infectious disease vectors, air pollution, etc

---

Many epidemiological studies have implicated solar radiation as a cause of skin cancer in fair-skinned humans. Recent assessments by the United Nations Environment Program project increases in skin cancer incidence and sun burn severity due to stratospheric ozone depletion for at least the first half of the twenty-first century. The groups most vulnerable to skin cancer are white Caucasians, especially those of Celtic descent living in areas of high ambient UVR. Further, culturally-based behavioural changes have led to much higher UV exposure, through sun-bathing and skin tanning.

The marked increase in skin cancers in western populations over recent decades reflects, predominantly, the combination of background, post-migration, geographical vulnerability and modern behaviours. Scientists expect the combined effect of recent stratospheric ozone depletion and its continuation over the next 1-2 decades to be (via the cumulation of additional UVB exposure), an increase in skin cancer incidence in fair-skinned populations living at mid to high latitudes. The modelling of future ozone levels and UVR exposures study has estimated that, in consequence, a 'European' population living at around 45 degrees North will experience, by 2050, an approximate 5% excess of total skin cancer incidence.

The equivalent estimation for the US population is for a 10% increase in skin cancer incidence by around 2050. Laboratory studies demonstrate that exposure to UVR, in particular to UVB, in various mammalian species induces lens opacification. The epidemiological evidence for a role of UVR in human lens opacities is mixed. Cataracts are more common in some (but not all) countries with high UVR levels. In humans and experimental animals, UVR exposure, including within the ambient environmental range, causes both localised and whole-body immunosuppression.

UVR-induced immunosuppression could influence patterns of infectious disease. It may also influence the occurrence and progression of various auto immune diseases and less certainly, vaccin efficacy. Finally, there is a wider, ecological, dimension to consider. Ultraviolet radiation impairs the molecular chemistry of photosynthesis both on land (terrestrial plants) and at sea (phytoplankton). This could affect world food production, at least marginally, and thus contribute to nutritional and health problems in food-insecure populations. However, as yet there is little information about this less direct impact pathway.

Health Impact Assessment (HIA)

Health impact assessment (HIA) has been defined as "a combination of procedures, methods and tools by which a policy, project or hazard may be judged as to its potential effects on the health of a population, and the distribution of those effects within the population". Despite recent advances in health impact assessment methods, its integration into mainstream policy-making has yet to be satisfactorily achieved. Besides, impact assessments typically refer to health impacts over the next 10 to 20 years (e.g. due to current smoking rates, obesity levels, or population ageing), rather than the 50 to 100 year time-scale appropriate to climate change projections. So there is need for scenario-based impact assessments that incorporate, and communicate, a higher level of uncertainty.

Several types of health impact assessments have been undertaken. A basic assessment identifies the types, but not much about the magnitudes, of potential impacts. In contrast, comprehensive well-funded and well-supported assessments are undertaken. For example, in the United States assessment, published in 2000, population health was one of the five target sectors included in the 16 detailed regional assessments and in the overall assessment. The US assessment involved stakeholder participation and extensive consultation and peer review.

Comprehensive multi-sectoral assessments have been conducted by the USA, Canada, the UK and Portugal. Assessments in developing countries have been undertaken only under the auspices of donor-funded capacity-building initiatives. The outcomes listed refer to the likely health impacts reported on for that particular country. The level of uncertainty accompanying these estimates is usually not described. Vector-borne diseases, particularly malaria, have been widely addressed. Other potentially greater impacts, such as from weather disasters, have been less well addressed. Out of these experiences, several conclusions can be drawn:

— Assessments should be driven by region and country priorities in order to determine which health impacts are considered. No single set of guidelines covers all health and institutional situations.

— HIA is a policy tool, therefore the actual process of conducting assessments, particularly the involvement of stakeholders, is very important.

— Assessments should set an agenda for future research. Nearly all the assessments done to date have identified research gaps, and they often specify detailed research questions.

— Assessment should be linked to follow-up activities such as monitoring and updated reports.

The development of formal guidelines for the national assessment of health impacts will improve methods used, will achieve some standardisation, and will facilitate the development of relevant indicators. Health Canada has prepared an initial framework, proposing that there are three distinct phases to the assessment task:

1. *Scoping*: to identify the climate change problem (concerns of vulnerable groups) and its context, describe the current situation (health burdens and risks) and identify key partners and issues for the assessment.

2. *Assessment:* estimations of future impacts and adaptive capacity, and evaluation of adaptation plans, policies and programmes.

3. *Risk management:* actions to minimise the impacts on health, including follow-up assessments.

This type of health impact assessment, in relation to largescale climatic-environmental changes, requires guidelines that accord with the mainstream HIA framework of WHO and other international agencies. Achieving this would help to move the climate change policy discussion beyond the environmental impact domain and into the social and public health impacts arenas. Currently, inmost countries, sector differentiation and the associated policy environment neither facilitates nor fosters intersectoral collaboration.

Within the health sector, resources are allocated primarily in relation to dealing with existing problems, taking some account of the relative burden of disease. A major shortcoming of many climate change health impact assessments has been the superficial treatment of the population's adaptive capacities and policy options. Strategies to enhance population adaptation should promote measures that are not only appropriate for current conditions, but which also build the capacity to identify and respond to unexpected future stresses/hazards.

The restoration and improvement of general public health infrastructure will reduce population vulnerability to the health impacts of climate change. In the longer-term, and more fundamentally, improvements in the social and material conditions of life and the reduction of inequalities within and between populations are required for sustained reduction in vulnerability to global environmental change.

Monitoring the Health Effects

Good evidence requires good data. The climate varies naturally as well as in response to human influences, and, in turn, climate is only one of many determinants of population health. Therefore, assessing the health impacts of climate change poses challenges. Further, the process of climate change is detectable only over decades, and the resultant health impacts will be similarly slow to emerge. Monitoring is "the performance and analysis of routine measurements aimed at detecting changes in the environment or health of populations".

In many public health investigations, it is possible to measure changes in a defined health impact and to attribute this trend to changes in a directly-acting risk factor. However, the monitoring of the impacts of climate change on health is more complex. There are three main issues:

(i) *Distinguishing apparent from real "climate change":* Climate is always fluctuating naturally, and many indices of health show seasonal and interannual fluctuation. The demonstration of such a relationship provides no direct evidence that climate change per se has occurred—rather, it merely confirms that these diseases have a seasonal or climatic dependence. An excess of heat-related deaths in a particularly hot summer, or even a succession of hot summers, indicates the potential for climate change to increase mortality, but it does not prove that mortality has increased as a result of climate change. That would require evidence of a change in the 'baseline' climate conditions—i.e. that the sequence of hot summers was exceptional, and due to climate change rather than random variation.

(ii) *Attribution:* Since climate is one of many influences on health, the attribution of an observed change in population health to an associated change in climate is not straightforward. The influence of concurrent changes in other environmental, social or behavioural factors must be first allowed for.

(iii) *Effect modification:* Over time, as the climate changes, other changes may also occur that alter the population's vulnerability to meteorological influences. For example, vulnerability to extreme weather events, including floods and storms, will depend on where and how residential housing is built, what flood protection measures are introduced, and how land-use is changed. Effective monitoring must include parallel measurements of population and environmental data, to allow study of potential modifying influences.

The principal criteria for selecting diseases and settings for monitoring should include the following:

— *Evidence of climate sensitivity*: to be demonstrated through either observed health effects of temporal or geographical climate variation, or evidence of climate effects on components of the disease transmission process in the field or laboratory.

— *Significant public health burden*: monitoring should be preferentially targeted towards significant threats to public health. These may be diseases with a high current prevalence and/or severity, or considered likely to become prevalent under conditions of climate change.

— *Practicality*: logistical considerations are important given that monitoring requires dependable and consistent long-term recording of health-related indices and other environmental parameters. Monitoring sites should be chosen where change is most likely to occur, but where appropriate capacity for reliable measurement exists.

The data needed for monitoring climate effects on health comprise: (i) climatic variables; (ii) population health markers; and (iii) other nonclimatic explanatory factors.

The choice of non-climatic variables will depend on the specific disease, but the principal categories of confounding or modifying factors include:

— age structure of population

— underlying rates of disease, especially cardiovascular and respiratory disease and diarrhoeal illness

— level of socio-economic development

— environmental conditions, e.g. land-use, air quality, housing conditions

— quality of health-care

— specific control measures, e.g. vector control programmes.

Specific Categories of Health Impacts

To monitor the health effects of thermal extremes, reliable longtime-series of temperature and mortality/morbidity data are available in many countries. An important focus of research data should be the assessment of how the temperature-mortality/morbidity relationship is modified by individual, social and environmental factors. Existing databases (e.g. EMDAT) for extreme weather events may be a key resource. To maximise their usefulness, complete and consistent reporting of extreme weather events across a wide geographical area, along with standard definitions of events and methods of attribution, is needed.

Current monitoring data can provide only a broad quantification of the relationship between climate and most vector-borne disease. Assessment of the climate contribution to long-term trends requires linked data on factors such as land-use, host abundance and intervention measures. Clearer understanding of relationships should result from high-quality serial data on vectors at a modest number of sites within or at the margins of endemic areas. Data from sites along specified transects could indicate changing vector distributions. Geographical comparisons based on remote sensing data may give additional insights into disease trends.

Adaptation and Adaptive Capacity

The IPCC has defined the following two closely-related terms:

1. Adaptation and Prevention
2. Adaptive Capacity

Adaptation and Prevention

Many adaptive measures have benefits beyond those associated with climate change. The rebuilding and maintaining of public health infrastructure is often viewed as the "most important, cost-effective and urgently needed" adaptation strategy. This includes public health training, more effective surveillance and emergency response systems, and sustainable prevention and control programmes. Extreme weather events can have vastly different impacts because of differences in the target population' scoping capacity. For example, cyclones in Bangladesh in 1970 and 1991 are estimated to have caused 300,000 and 139,000 deaths respectively.

In contrast, Hurricane Andrew struck the United States in 1992, causing 55 deaths (although also causing around $30 billion in damages). Climate-related adaptation strategies must therefore be considered in relation to broader characteristics—such as population growth, poverty, sanitation, healthcare, nutrition, and environmental degradation—that influence a population's vulnerability and capacity to adapt. Adaptations which enhance a population's coping ability may protect against current climatic variability as well as against future climatic changes. Such "no-regrets" adaptations may be especially important for less developed countries with little current coping capacity.

Adaptive Capacity

Adaptive capacity refers to both actual and potential features. Thus, it encompasses both current coping ability and the strategies that expand future coping ability. For example, access to clean water is part of the current coping capacity for developed countries—but represents potential adaptive capacity in many less developed countries. Highly-managed systems, such as agriculture and water resources in developed countries, are thought to be more adaptable than less managed or natural ecosystems.

Unfortunately, some components of public health systems are often relaxed when a particular health threat recedes. For example, the threat of infectious diseases appeared to be retreating thirty years ago because of advances in antibiotic drugs, vaccines and pesticides. Today, however, there is a general resurgence of infectious diseases—and relevant public health measures need to be reinvigorated. The main determinants of a community's adaptive capacity are: economic wealth, technology, information and skills, infrastructure, institutions, and equity. Adaptive capacity is also a function of current population health status and pre-existing disease burdens.

Economic resources

Wealthy nations are better able to adapt because they have the economic resources to invest, and to offset the costs of adaptation. In general, poverty enhances vulnerability—and we live in a world in which approximately one-fifth of the world's population lives on less than US$1 per day.

Technology

Access to technology in key sectors and settings is an important determinant of adaptive capacity. Many health protecting adaptive strategies involve technology—some of which is well established, some new and still being disseminated, and some still being developed to enhance coping with a changing climate. The health risks from proposed technological adaptations should be assessed in advance. For example, increased air conditioning would protect against heat stress, but could increase emissions of greenhouse gases and other air pollutants. Poorly designed coastal "defences" may increase vulnerability to tidal surges if they engender false security and promote low-lying coastal settlements.

Information and skills

In general, countries with more "human capital" or knowledge have greater adaptive capacity. Illiteracy increases a population's vulnerability to many problems. Health systems are labour-intensive and require qualified and experienced staff, including those trained in the operation, quality control, and maintenance of public health infrastructure.

Infrastructure

Infrastructure specifically designed to reduce vulnerability to climate variability (e.g., flood control structures, air conditioning, and building insulation) and general public health infrastructure (e.g., sanitation facilities, waste water treatment systems, laboratory buildings) enhance adaptive capacity. However, infrastructure (especially if immovable) can be adversely affected by climate, especially extreme events such as floods and hurricanes.

Institutions

Countries with weak institutional arrangements have less

adaptive capacity than countries with well established institutions. For example, institutional and managerial deficiencies contribute to Bangladesh's vulnerability to climate change. Collaboration between public and private sectors can enhance adaptive capacity. For example, the Medicines for Malaria Venture—a joint public-private initiative to develop new antimalarial drugs—is developing new products for use in developing countries.

Equity

Adaptive capacity is likely to be greater when access to resources within a community, nation, or the World is equitably distributed. Under-resourced and marginal populations lack adaptive resources. While universal access to quality services is fundamental to public health, many still lack access to health care. Overall, the developing world, with 10 per cent of the World's health resources, carries 90 per cent of the disease burden.

Health Status and Pre-existing Disease Burdens

Population well-being is an important ingredient and determinant of adaptive capacity. Great progress has been achieved in public health, yet 170 million children in poor countries are underweight, of whom over three million die each year. Many countries face the double burden of increases of non-communicable diseases, but with continued prevailing infectious diseases.

DEVELOPING RESPONSES TO CLIMATE CHANGE

To make informed decisions about climate change, policy-makers will need timely and useful information aoout the possible consequences of climate change, people's perceptions of those consequences, available adaptation options, and the benefits of slowing the rate of climate change. The challenge for researchers is to provide this information. Once policy-makers have received input from the impact assessment community, they must integrate this information into a broader policy portfolio.

Response options include actions to mitigate greenhouse gase missions to slow the rate of climate change; measures to adapt to a changing climate in order to increase society's resilience to the changes that are coming; activities to increase the public's awareness of the climate change issue; investments in monitoring and surveillance systems; and investments in research to reduce key policy-relevant uncertainties. Climate change, however, should not be considered in isolation from other global environmental stresses. Further, policy-makers usually deal with multiple social objectives (e.g., poverty elimination, promotion of economic growth, protection of cultural resources), while competing stakeholder desires compound the allocation of scarce resources.

Climate change should therefore be viewed as part of the larger challenge of sustainable development. Using the information provided by the research community, risk managers must make decisions despite the existence of scientific uncertainties. Policy-focused assessments analyse the best available scientific and socioeconomic information to answer questions being asked by risk managers. They characterise and, if possible, quantify scientific uncertainties to the extent possible, and explain the potential implications of the uncertainties for the outcomes of concern to the decision makers. Ultimately, it is upto society to decide whether a perceived risk warrants action. But the scientific uncertainty, by itself, does not excuse delay or inaction.

Decision-making Criteria

Many different criteria exist for making decisions about climate change policy. Two approaches to decision making that are often discussed are the "precautionary principle" and "benefit-cost" analysis. The precautionary principle is a risk management principle applied when a potentially serious risk exists, but significant scientific uncertainty also exists.

The precautionary principle allows some risks to be deemed unacceptable not because they have a high probability of occurring, but because the consequences if they occur may be severe or irreversible. This principle was featured in the 1992 Rio Declaration on Environment and Development as Principle 15, stating: "Where there are threats of serious or irreversible damage, lack of full scientific certainty shall not be used as a reason for postponing cost-effective measures to prevent environmental degradation."

Another widely used approach is the "benefit-cost" criterion, weighting the expected benefits and costs of a proposed action. Questions arise about how benefits and costs should be measured, and how they should be compared among different societies. The benefit-cost criterion emphasises the efficient use of scarce resources but does not deal with equity. Nor does it deal well with consequences that are displaced into the future, and therefore, by economic convention, often discounted. Climate change has the potential for catastrophic outcomes in the distant future, the "present value" of which would be small if discounted. Despite these concerns, benefit-cost analysis should not be dismissed. This would only deprive decision-makers of one set of insightful information.

The mitigation of greenhouse gases provides a mechanism for slowing, and perhaps eventually halting, the buildup of greenhouse gases in the atmosphere. A slowing of the rate of warming could yield important benefits in the form of reduced impacts to human health and other systems; however, the inertia in the climate system means that there will be a significant temporal lag between emission reduction and slowing in the rate of warming. Adaptation is another important response option. Such actions enhance the resilience of vulnerable systems, thereby reducing potential damages from climate change and climate variability.

Communication of information about climate change, its potential health impacts, and response strategies, is itself a public policy response to climate change. So, too, are the development and implementation of monitoring and surveillance systems, and investments in research. Monitoring and surveillance systems are integraland essential to providing the information needed to support decisions by public health officials.

Policy-focused Assessment

Policy-focused assessment is a process that can help resource managers and other decision-makers meet the challenge of assembling an effective policy portfolio. It is a process by which the best-available scientific information can be translated into terms that are meaningful to policymakers. A policy-focused assessment is more than just a synthesis of scientific information or an evaluation of the state of science. Rather, it involves the analysis of information from multiple disciplines—including the social and economic sciences—to answer the specific questions being asked by stakeholders. And it includes ananalysis of adaptation options to improve society's ability to respond effectively to risks and opportunities as they emerge.

Formulating good policy requires understanding the variability in vulnerability across population sub-groups, and the reasons for that variability. In the assessment of adaptation options, a number of factors related to the design and implementation of strategies need to be considered. These include the fact that

1. the appropriateness and effectiveness of adaptation options will vary by region and across demographic groups;
2. adaptation comes at a cost;
3. some strategies exist that would reduce risks posed by climate change, whether or not the effects of climate change are realised;
4. the systemic nature of climate impacts complicates the development of adaptation policy; and
5. maladaptation can result in negative effects that are as serious as the climate-induced effects being avoided.

Complicating the assessment process is the fact that there are significant scientific and socioeconomic uncertainties related to climate change and its potential consequences for human health. Uncertainties exist about the potential magnitude, timing and effects of climate change; the sensitivity of particular health outcomes to current climatic conditions (i.e., to weather, climate, and climate-induced changes in ecosystems); the future health status of potentially affected populations (in the absence of climate change); the effectiveness of different courses of action to adequately address the potential impacts; and the shape of future society (e.g., changes in socioeconomic and technological factors).

A challenge for assessors is to characterise the uncertainties and explain their implications for the questions of concern to the decision-makers and stakeholders. If uncertainty is not directly addressed as part of the analysis, a health impacts assessment can produce misleading results and possibly contribute to ill-informed decisions.

8

Effects of Climate Change on Coastal Ecosystems

The complex processes (e.g., solar input, regional heating and cooling, pressure gradients, circulation) that occur in and affect the atmosphere, sometimes called an "ocean of air," are similar in many ways to the complex processes that occur in and affect the oceans. Biological and chemical processes in the biosphere and geosphere affect the atmosphere and the oceans, and the atmosphere and oceans affect the biosphere and the geosphere as well as each other. The cryosphere influences the atmosphere, partly because its high albedo reflects radiation back into space, and the hydrosphere, partly because sea ice formation influences freshwater balance.

Our knowledge of the intricate, interacting processes in and among these spheres is rudimentary, which makes predictions about the extent and direction of global climate change a grand challenge. In addition, changes may be non-linear, so they are not readily predicted from what has been observed in the past, thereby resulting in surprises. Changes may be extreme, and they may be the result of a "flip," or rapid change, from one stable state to another.

Coastal zone is a region of land margins with dynamic interactions among atmospheric, terrestrial, and aquatic systems. These interactions result in high biological productivity and resources of important economic value. For example, the coasts of the United States and its territories extend for about 80,000 miles and include the coasts of the continental United States as well as Alaska, Puerto Rico, the Pacific islands of Hawaii, American Samoa, and Guam, and a number of smaller island groups in the Pacific and Caribbean.

The narrow coastal strip of the contiguous United States represents 17 percent of the nation's continental land area, but is inhabited by about 53 percent of the population. These are the most heavily developed regions of the nation, with rates of recent population growth that exceed that of the country as a whole. This is important because coastal systems cannot be considered without reference to what is happening on the adjacent land mass.

Material washing off land or delivered from the atmosphere with precipitation enters estuaries or the sea, and nutrient and contaminant runoff from densely populated regions stresses coastal ecosystems. Within the coastal zone, estuaries and their associated wetlands are vital to the well-being of coastal systems and fisheries, serving not only as harvest sites, but also as nurseries for commercial species harvested elsewhere. Offshore, the U.S. Exclusive Economic Zone beyond and adjacent to the territorial sea is estimated to be habitat for about 20 percent of all harvestable seafood in the world.

Coastal and offshore systems are currently under heavy harvest pressure, which may threaten not only the fisheries, but also the very existence of some harvested species. Because those species are part of a larger web of feeding and other biological relationships, their depletion or loss may have important negative consequences for the ecosystems they inhabit and the reptiles, birds, and mammals that feed on them. If climate change adds stress to coastal and offshore systems, that may have major negative effects on some ecosystem components. This situation is analogous to that faced by a human whose immune system is compromised and who may succumb to a disease that would not threaten a healthy person.

Climate Change on Coastal Systems

In terms of global climate change, environmental factors that are expected to have the greatest direct effects on estuarine and marine systems are temperature change, sea-level rise, availability of water from precipitation and runoff, wind patterns, and storminess. The state of our knowledge of these environmental factors allows us to make reasonable predictions about some of their effects, but not about others. Temperature influences organism biology (mortality, reproduction, growth, behaviour), affects dissolved oxygen concentrations in water, and plays a direct role in sea-level rise and in major patterns of coastal and oceanic circulation.

There is some confidence about predicting what would happen to organisms, oxygen concentrations, and sea levels as temperature increases, less confidence as to temperature's influence on interactions among organisms (e.g., predator-prey, parasite-host, competition for resources), and even less as to its effect on water circulation patterns. There is some confidence about predicting the effects of sea-level rise on shallow continental margins, including flooding of wetlands, shoreline erosion, and enhanced storm surges. There is less confidence in the ability to predict the effects of climate change on precipitation, wind patterns, and the frequency and intensity of storms.

Precipitation affects runoff into estuaries and therefore influences estuarine circulation, concentrations of nutrients and contaminants, stratification and oxygen deficits, and recruitment of some species (i.e., the addition of new individuals to a species' population). Wind speed and direction influence coastal circulation, including currents that deliver fish and invertebrate larvae from coastal waters into estuaries. Storms and their associated winds can have major negative effects on coastal ecosystems and shoreline structures.

Changes in Environmental Factors

Temperature

Temperature extremes can be lethal to organisms. Because evaporative cooling limits the level to which water temperatures can increase, summer temperatures of tropical waters may not become much higher than they are now as the climate warms. However, temperatures in temperate and boreal regions may rise to levels that are stressful or lethal to resident organisms. Many species are sensitive to temperatures just a few degrees higher than those they usually experience in nature. Indeed, in short-term temperature-tolerance experiments on estuarine invertebrates in the laboratory, a temperature increase of 1°C often raised mortalities from ~0 percent at the lower temperature to ~100 percent at the increased temperature.

Temperature increases in the coming century are predicted to be greater than 1°C. At sub-lethal levels, temperature influences growth and metabolism, governs animal behaviour and distribution patterns of organisms, and acts in concert with other environmental variables such as dissolved oxygen. It influences the timing of reproduction and controls rates of egg and larval development. For example, recent declines in the abundance of winter flounder in New England might be due to the negative effects of warmer winters on egg survival and larval hatching and development.

Higher temperatures also may have positive effects. Some commercially valuable estuarine-dependent species in the lower latitudes have higher growth rates and larger annual harvests when temperatures are higher. For example, shrimp harvest is generally highest in tropical climates and declines northward, so an increase in temperatures could increase the annual yield of shrimp in temperate waters like the Gulf of Mexico, as long as temperatures did not exceed lethal levels. In the mid-Atlantic region, where severe winters are thought to result in low blue crab catches, less severe winters may enhance harvests if other factors do not intervene.

Temperature directly or indirectly influences population abundances and distributions of organisms. For example, temperature cycles and maximum abundances have been strongly correlated for over 30 commercial fish species in the Gulf of Maine, and inshore sea-surface warming and offshore cooling was associated with a period of exceptional fish production on the North American West Coast. The distributional limits of southern species in New England moved north as water temperatures increased north of Cape Cod after the 1940s, and marine temperature variation explained changes in the north-south distribution of 12 of 36 species of fish in the U.S. mid-Atlantic region over about 20 years.

Higher winter temperatures on a North Carolina inshore reef were associated with greater numbers of 28 species of tropical fish on the reef over a 15-year period. Similar climate-influenced changes would have positive effects on some fisheries and negative effects on others, presenting a challenge to managers of and participants in particular fisheries. Although higher temperatures might not result in extinction of a species throughout its range, the species may be eliminated from part of its range.

Temperature changes as small as 1°C can have important effects on species distributions in a short time. For example, the number of species of reef fish off Los Angeles declined by 15 to 25 percent between 1976 and 1977 when temperatures increased abruptly by ~1°C, demonstrating that regional species diversity of mobile organisms can change rapidly. Temperatures in the region remained higher through the 1990s and the proportion of southern species rose while the proportion of northern species fell. As another example, the annual mean water temperature inshore at Monterey Bay, California, increased by ~0.8°C and mean summer maximum temperatures were ~1.9°C higher between the early 1930s and the late 1990s.

By 1996, the abundances of 10 of 11 intertidal species with southern distributions were higher than their abundances in the early 1930s, whereas the abundances of 5 of 7 northern species declined in the interim. Such examples of changed patterns of distribution in concert with warming allow us to develop scenarios of changes in U.S. coastal regions. Thus, species whose southern range ends in the mid-Atlantic region will be lost from that region if water temperatures reach levels that are lethal or that inhibit successful reproduction and recruitment. For example, the commercially important soft clam in the Chesapeake Bay lives near its southern distribution limit and may be eliminated from the bay if bottom water temperatures approach and remain near 32°C in summer.

Similarly, the commercially important ocean quahog clam lives near its southern limit in the mid-Atlantic region. Weinberg et al. estimate that an increase of 2°C would extend the period of thermal stress for ocean quahog larvae from the present 167 days to 231 days, lowering survival of the larvae and depressing their recruitment to the resident population, which could eventually die out over a period of decades. Unfortunately, there are few data on temperature tolerances available for most other species in the mid-Atlantic region, so data on geographic ranges must be used to predict warmingrelated shifts in distribution.

Fish communities in the mid-Atlantic region contain mainly species with either northern or warm-temperate distributions, so there should be changes in fish distributions if the climate warms. Based on distribution information in Murdy et al., the southern range limits of cooler-water fish like winter flounder and cunner would retreat northward, whereas warmer-water species like southern flounder and spotted seatrout would range even farther north than they do now. Similar predictions about expanded distributions northward can be made for some mid-Atlantic invertebrates. For example, pink, white, and brown shrimp support commercial fisheries as far north as North Carolina and range north of Cape Hatteras in noncommercial abundances.

If temperature is the factor limiting their northward distribution, climate warming should allow them to become established in the mid-Atlantic region and in the lower Chesapeake Bay in numbers that might support a fishery. Based on distribution data in Emmett et al., similar predictions can be made for Pacific coast organisms. Thus the southern range boundaries of starry flounder, the Dungeness crab, the soft clam, and the fat gaper clam may retreat northward as climate warms while the northern range boundaries of Pacific hake, some bass in the genus Paralabrax, and the Pacific gaper clam may advance northward.

Movement into and colonisation of newly suitable habitat depends on many factors:

1. the number of adults available in the original habitat and their ability to produce young (seeds, spores, larvae);
2. an adequate number of potential colonisers (seeds, spores, larvae, migrating juveniles or adults);
3. the ability of potential colonisers to move into the new habitat, including their ability to cross barriers; and
4. the survival of adequate numbers of individuals in the new habitat to ensure genetic diversity to meet environmental challenges and to produce succeeding generations.

Corridors link a region where a species is at risk and a more suitable region. Such corridors may not exist for estuarine species that cannot survive the high-salinity marine conditions that occur between estuaries. Thus, the loss of species from an estuary that has become too warm for those species may reduce species diversity in that estuary in the short term, depending on the speed of immigration to that estuary of species that can tolerate the warmer conditions. This is especially true of species that have limited abilities to disperse or where there are no dispersal corridors.

Mobile organisms like fish, swimming crabs, and shrimp can quickly colonise new habitats by migrating along the coast, but immobile or relatively immobile invertebrates such as clams and oysters will migrate more slowly. Coastal plants may also be sensitive to changes in air and water temperatures. For example, changes in frequency of freeze events will alter ecosystems of the east and west coasts of the Gulf of Mexico because shoreline vegetation is strongly influenced by the frequency of frost from Texas to south Florida.

In south Florida, episodic frost or freezing events help shape ecosystems because tropical plants and animals are highly vulnerable to such events. If climate change reduces the frequency of episodic freeze events, over time the coastal red mangrove communities should shift farther north of Tampa Bay on Florida's Gulf of Mexico coast and Laguna Madre on the Texas coast. Greater development of black mangrove forests, whose current infrequent occurrence is associated with the higher frequency of frost, would be expected along the Louisiana coast. Also, subtropical seagrasses would be more prevalent in warm-temperate coastal habitat where suitable water clarity exists.

Temperatures can also influence changes in species interactions in ecosystems. For example, winter flounder may be negatively affected by the extent of warming predicted for New England. As noted earlier, their abundances in southern New England have been declining but the reasons are unclear. An alternative hypothesis to that of Keller and Klein-MacPhee is that the decline is due to a temperature-mediated shift in the food web that favours organisms that feed in the water column, thus reducing the amount of food available to bottom feeders like winter flounder. Changes in distribution patterns would change the mix of predators, prey, parasites, and competitors in an ecosystem in ways that are not yet understood but that could alter the functions of the ecosystem and the productivity of selected fisheries.

Predation pressure in marine ecosystems generally increases from the poles to the tropics, so warming due to climate change could cause an ecological shift to increased predation if it led to greater diversity and numbers among predators. For example, warm winter conditions of the past decade have favoured predatory game fishes that have become more numerous in the Laguna Madre than elsewhere in Texas. In turn, densities of small fishes, shrimps and crabs are lower in Laguna

Madre seagrass beds than elsewhere in the Gulf of Mexico. Climate change may also alter species interactions by changing the timing of physiological events. For example, altered temperature regimes could advance or retard the timing of reproduction for many species.

Where the production of young of a species is synchronised with the presence of their food supply, temperature-induced changes in production of young may result in their being present when their usual food supply, if it does not respond to changed temperature, has not developed or has passed its peak. For example, such "synchronicities" of timing affect migrating waterfowl and shorebirds that depend on the short-term population explosion of invertebrate prey to support food requirements of adults and young on their high-latitude breeding and rearing grounds.

Temperature influences the amount of oxygen that water can hold. Most aquatic organisms extract the oxygen necessary for their metabolic demands and survival from the water in which they live. Interactions between higher temperatures and depleted oxygen could constrict the available habitat for certain aquatic species. For example, striped bass habitat may be "squeezed" along the North American East Coast, especially in Chesapeake Bay, an important spawning and rearing region for this species.

The "squeeze" can occur because the surface waters of the striped bass' normal habitat may become so warm that the fish will seek cooler waters. However, the cooler, deeper waters will not contain enough oxygen if hypoxia or anoxia is present and will be avoided by the fish. Laboratory studies have shown that organisms under stress pay a metabolic cost in the form of a continued expenditure of energy that may lead to death if the stress does not abate.

Increased ocean temperatures may alter coastal ocean currents. For example, Gallegos et al. propose that global climate-induced changes in ocean temperatures might alter circulation patterns of the Gulf of Mexico and Caribbean. This has important implications for coastal south Florida, where the major circulation systems influence the distribution, recruitment, and survival of coastal marine fish and invertebrate communities of the Florida Keys and other areas of south Florida. These coastal boundary currents also influence the residence time of water in nearshore environments and serve as transport mechanisms for materials along the Gulf of Mexico coast. Many recreationally and commercially important fish and shellfish species in the Gulf spawn offshore, especially in winter, but their juveniles depend on coastal and estuarine nursery areas. Alteration of coastal currents and offshore-inshore transport corridors may have negative consequences on the growth and survival of these species.

Sea-level rise

Climate warming will lead to thermal expansion of water and melting of glacier and polar land ice, with a subsequent rise in sea level. The effects of sea-level rise will vary with location, speed of the rise, and the geological and biological responses of the affected ecosystems. Sea-level rise is a particular threat to the low-lying, shallow-gradient wetlands of the Middle and South Atlantic and Gulf of Mexico. Coastal storms put the two Atlantic regions at further risk from sea-level rise, and the central Gulf Coast region faces an added risk because the deltaic plain is sinking. In general, sea-level rise will inundate coastal lands and erode susceptible shores.

In salt marsh and mangrove habitats, sea-level rise may submerge wetlands, waterlog soils, and cause plant death from salt stress. Most wetland habitats can survive sea-level rise by migrating inland to areas of decreasing tidal inundation along undeveloped shores with relatively gentle slopes. In the absence of human effects, there would not be a catastrophic effect on most Gulf of Mexico coastal wetlands, at least at the rates of rise currently projected for the next 50 to 100 years.

On developed coasts, wetland plant systems cannot move inland as sea level rises on sedimentary shores because seaside development by humans limits that option. If sediment inputs are limited or prevented by the presence of flood-control, navigational, or other anthropogenic structures, marshes and mangroves may be starved for sediment and normal accretion will be hindered. If inundation outpaces accretion, the marsh or mangrove forest will be submerged, die, and disappear.

The continued loss of wetlands in Louisiana indicates that, under the present rate of sea-level rise of 2.3 mm per year and land subsidence of 4 mm per year, total marsh area will be critically reduced as humans alter water and sediment distributions. Estimates of the critical rates of sea-level rise above which south Florida and Caribbean mangrove ecosystems will collapse range from 1.2 mm per year to 2.3 mm per year. However, these rates are only for mangroves in regions with little sediment input, and there has been strong criticism of these threshold values.

The most vulnerable mangrove ecosystems in the Gulf of Mexico include those in micro-tidal limestone settings, and those with extensive inland residential development. For example, the extensive coastal development of south Florida would squeeze many species into smaller habitat spaces, with high risk to vulnerable species such as the endangered American crocodile and perhaps sea turtles. Louisiana is already undergoing high rates of relative sea-level rise because of land subsidence.

Development of interior wetland margins by humans as well as the relatively steep gradient of the Pleistocene terrace escarpment prevent landward migration of most Louisiana wetlands, so this ecosystem can be maintained only through vertical accumulation of soil. This process is impaired in Louisiana by major human modifications to the hydrology of the Mississippi River deltaic plain. However, where the Mississippi River connects with its deltaic plain, such as in western Terrebonne Bay where the outflow from the Atchafalaya River nourishes adjacent areas, coastal wetlands appear to respond by adequately maintaining their elevation in the face of subsidence.

The same is true for largely unaltered coastal basins such as Pascagoula Bay in Mississippi. It is when sea-level rise is combined with other human land use changes in the drainage basins of the Gulf of Mexico that the consequences may be severe. Wetland marshes are important contributors to the biological productivity of coastal systems and function as nurseries and as refuges from predators for commercially important shellfish and fish and for birds like some rails, gulls, and terns.

Rising sea levels will initially increase access to marsh surfaces by fish and invertebrates, perhaps increasing their production in the short term. However, depletion or loss of marshes will have important effects on nutrient flux, energy flows, essential habitat for a multitude of species, and biodiversity. Such effects will be positive or negative, depending on the species. For example, birds that require the marsh for rearing young will be affected negatively by its loss, whereas birds that feed in shallow water or on intertidal sand and mud flats that replace the marsh will be affected positively.

Some coastal zones are habitat for rare, threatened, or endangered species of plants and animals and, if the changes threaten rather than enhance their habitat, they may be lost. Such loss represents the loss of much unique and potentially significant genetic information. Human interference with natural barrier island processes increases the islands' vulnerability to even a small rise in sea level. Many barrier islands move toward the mainland through a process of beach erosion on their seaward flank, overwash of sediment across the island during storms, and deposition of the eroded sediment in the quieter waters of the inland bay. The rate of this natural migration depends largely on the rate of sea-level rise, and also on the frequency and severity of storms and hurricanes.

Even a small rise in sea level sets this process in motion. The construction of buildings, roads, bulkheads, seawalls, and any other type of shore hardening disrupts overwash and sediment movement. In response, the beach face and nearshore erode, threatening buildings and narrowing the beach, but the "ribbon of sand" does not grow landward. Eventually, this threatens not only human habitation on barrier islands but also the barrier islands themselves and the habitats they provide.

Finally, accelerated rates of subsidence and sea-level rise may alter the depth and width of shallow estuaries. For example, many estuaries in the Gulf of Mexico are only 1 to 3 meters deep and have little topographical relief, so that small absolute changes in relative sea level will greatly change their relative bottom area. Because of the strongly wind-driven nature of these estuaries, bottom friction is very important in controlling hydrodynamics. Changes in bottom area will have an immediate and important effect on the pattern of energy dissipation in estuaries. Rapid changes in bottom area and configuration can also have ecological consequences if they bury or scour important fishery nursery habitats such as seagrass meadows and oyster reefs.

Precipitation and hydrologic regime

Changes in precipitation and hydrologic regimes at the regional level can have important ecological effects on coastal ecosystems. Decreased rainfall and runoff combined with sea-level rise would result in salinity encroachment into the tidal freshwater reaches of an estuary, affecting lower-salinity communities of organisms. Decreased freshwater delivery alters food webs in estuaries and changes the residence time of nutrients and contaminants. Reduced nutrient input during drought can be a major cause of the loss of productivity of a river-dominated estuary during and after drought periods.

Permanent reductions of freshwater flows due to combined effects of human activities and climate change could lead to major reductions of biological productivity in alluvial bay systems such as the lagoons of Texas, Mobile Bay, Apalachicola Bay, and Tampa Bay. Whereas reduced river flow would reduce nutrient input in estuaries with rather pristine watersheds, there could be different effects in polluted watersheds. Watersheds next to heavily industrialised or urbanised estuaries have sources of nutrients and contaminants that are not a function of river flow.

Many of these estuaries also receive substantial amounts of nutrients such as nitrogen precipitated from the atmosphere. The combined effect of human development, nutrients transported by atmospheric processes, and reduced river flow would produce water quality conditions that would negatively affect fisheries and human health through such changes as accumulation of metals in animals and plants. Climate change-induced shifts in the recharge of aquifers together with increasing demand by humans on water resources could have important implications for coastal ecosystems. For example, a significant influx of fresh water into several estuaries and bays in the Gulf of Mexico comes from groundwater.

However, increased upstream consumption of water by agricultural, urban, and industrial users over the last two decades has reduced artesian outflows of groundwater downstream. The combination of increased demand and decreased recharge of this resource as a result of climate change would reduce estuarine productivity in this region and others where estuaries depend on ground water.

Although the occasional drought year has had a positive effect on the secondary productivity of estuaries in the Gulf of Mexico region, increases in drought intensity or frequency would increase the incidence of coastal hypersalinity, resulting in the decline of valuable habitats such as mangroves, seagrasses, and coral reefs. For example, the release of high salinity waters through the passes of the Florida Keys adversely affects coral reefs offshore.

Increased salinity in Florida Bay causes extensive die-offs of seagrasses and stress to mangroves on reef islands. Hypersalinity of south Texas lagoons during an extensive drought in the 1950s contributed to seagrass die-offs and fish kills, particularly in Laguna Madre, where salinities of 80 parts per thousand occurred. Even in the more humid central sector of the Gulf Coastal Plain, a 25-month drought is considered to be a primary cause of severe dieback of about 100,000 acres of salt marshes along coastal Louisiana during the summer of 2000.

Increased rainfall and resultant freshwater runoff into an estuary would shrink estuarine habitat if, at the same time, sea-level rise were moving seawater up an estuary where humans were preventing overflow onto land. This would also increase stratification of the water column, leading to sub-surface hypoxia or anoxia in estuaries with excess nutrients. If mid-Atlantic estuaries freshened sufficiently, abundances of the sea nettle, a jellyfish species that is negatively affected by low-salinity conditions, would decline. Such a decline in numbers might be ecologically significant because the sea nettle is an important predator of many other species in the food webs of the Chesapeake Bay.

Climate change may increase the frequency of intense precipitation events and thereby alter coastal ecosystems, depending on the degree of regional water management. For example, the intra- and inter-annual variability of precipitation has been a major force shaping many coastal ecosystems in the Gulf of Mexico. An increase in the intensity of precipitation will change the pattern of freshwater runoff in the coastal plain watersheds. Increased flooding, however, depends on how regional hydrology is managed. For those Gulf of Mexico estuaries that receive much greater inputs of freshwater inflows than they did historically, there has been a major shift in their ecological communities to those that are better adapted to lower salinities.

By contrast, estuaries and lagoons receiving much lower freshwater inflows than they did historically have shifted to communities better adapted to high-salinity conditions. Most coastal systems in the Gulf of Mexico would probably experience little direct effect from increased freshwater discharge, except that the estuaries that would receive most of the excess surface water from management canals could experience severe reductions in salinity. Thus, there is potential for adverse consequences. But, again, this depends at least as much on how the water system is managed as on a climate change-induced increase in precipitation.

With less precipitation, runoff will decline proportionately in many of the managed systems. The key point is that Gulf of Mexico estuaries may benefit from any degree of increased precipitation, but only a slight decrease in freshwater input may cause severe changes in habitat and water quality. This may be especially evident in regions of large engineered water management systems, or coastal systems in semi-arid conditions.

Wind, water circulation, and storminess

The ocean circulation can be divided into two parts—the thermohaline and the wind-driven circulations. The drag created by wind stress acting on the ocean's surface drives the major ocean surface currents. The winds themselves are created by the unequal heating of the earth's surface. Under most global warming scenarios, polar regions will experience higher temperatures that will reduce the thermal gradient between the poles and equator, which may result in a decrease in wind stress. This would potentially lead to a weakening of the overall wind-driven circulation.

Weaker wind-driven surface currents could seriously affect the structure and function of open ocean and nearshore ecosystems. As an example, the surface waters off the U.S. West Coast can be divided into three oceanographic domains: a coastal upwelling domain off California, a coastal downwelling domain off the southern coast of Alaska, and the Alaska gyre in the Gulf of Alaska. All three of these domains are organised by seasonally varying winds associated with overlying atmospheric pressure cells. Relaxation of the winds would reduce upwelling off California and within the Alaska gyre and reduce downwelling along the coast of Alaska.

Transport of nutrients and primary and secondary production would also be reduced. Wind speed and direction influence productivity of estuarine and marine systems. For example, nutrients that are delivered from deeper waters to the sunlit surface by the upwelling of water support phytoplankton production along coasts. Such upwelling is the result of winds that blow alongshore, and it can be diminished if the water is stratified by differences in temperature or salinity. If upwelling is slowed by changes in wind and temperature, phytoplankton production will be lowered.

Upwelling might continue from shallower waters, but there are fewer nutrients in these waters than in the deeper waters that are normally upwelled. The potential results of decreased upwelling are demonstrated by the 70 percent decline in zooplankton abundances off the California coast between 1951 and 1993 in concert with a warming of the sea surface (up to 1.5°C), which increased stratification and hindered upwelling. Hill summarises the apparent negative effects of these depleted zooplankton abundances on other members of the coastal food webs like fish and birds in the region. There has been some speculation that global warming may result in intensified coastal upwelling in some circumstances.

Most greenhouse gas models project more warming over land than over open oceans, resulting in an increased temperature contrast between land and ocean. This increased contrast would strengthen low-pressure cells that typically occur over land adjacent to offshore high-pressure cells. The relative strength and position of these pressure cells results in enhanced alongshore winds that are the driving force behind upwelling. On the U.S. East Coast, alongshore wind stress and differences in the densities, and therefore buoyancies, of fresher and saltier waters help produce water movements that transport larval blue crabs, menhaden, and bluefish in the Middle and South Atlantic Bights.

Climate-related changes in these circulation patterns that hinder such transport may lower abundances of these species within estuaries and along the coast, thereby modifying coastal ecosystems and affecting fishery yields. Winds may also affect fish production by their role in the large variability that occurs in fish recruitment. A growing consensus suggests that there is an "optimal environmental window" within which larval survival is maximised and outside of which survival is sharply curtailed. Applying this idea to fish recruitment in upwelling regions, Cury and Roy demonstrated that recruitment is maximised at wind speeds of roughly 5 to 6 meters per second. At lower wind speeds, vertical mixing is weak and food production is low.

At higher wind speeds, turbulence in the water column acts to disaggregate food and larvae patches. There is much uncertainty as to whether coastal storms and hurricanes will increase in number or intensity, or both. Any increase in the number and severity of coastal storms and storm surges could have serious implications for the well-being of fishery and aquaculture industries. An example of problems beyond damage to infrastructure is the deleterious effects of Hurricanes Dennis, Floyd, and Irene on Pamlico Sound, North Carolina, between September and October 1999. About 1 meter of rain from the storms led to 50- to 500-year floods that washed the equivalent of 6-months' supply of nitrogen into the sound.

Strong stratification of the water column ensued, bottom waters became hypoxic, algal blooms intensified, and fish diseases became more prevalent. The sound takes over a year to flush, so the effects of these hurricanes on ecosystem functions may be prolonged. The consequences of sea-level rise become acute during storm events because sea-level rise and storms interact to erode beaches and produce high storm surges.

Hurricanes and sea-level rise have shaped the coastal topography of the Gulf of Mexico for the last several thousand years. Although it is likely that some Gulf coastal ecosystems would not be damaged by present and even future rates of sea-level rise under average conditions, episodic events like hurricanes and fires may undermine their resilience. Where excessive wind damage stresses mangrove forests and limits their ability to produce peat, forest elevation cannot keep pace with sea level, so additional flooding results in plant mortality and irreversible loss of habitat.

If the future frequency of hurricanes were to stay the same as in the past, the ecological consequences of sea-level rise, together with increased coastal development, would be expected to inhibit natural recovery. More frequent storms and hurricanes might lead to increased local damage to mangrove forests, transient increases of sediment and organic load to coastal waters, and increased physical damage to coral reefs.

Influence of Coastal Aquaculture

Coastal aquaculture has become an important food source, so an understanding of the potential effects of climate change on cultured species is important to the industry. Global aquaculture production has increased since the 1950s, attaining a growth rate of ~10 percent annually since 1990; by 2030, aquaculture harvests are expected to be greater than capture harvests. Marine aquaculture production in the United States is expected to continue to grow as well. Higher temperatures could enhance growth rates of cultured species and allow for the culture of species in areas that are currently too cold for them.

On the other hand, a region may become too warm to allow the culture of a heat-sensitive species. For example, high temperatures are associated with summer mortality syndrome in cultivated Pacific oysters on the U.S.

West Coast, so increased warming may add to such mortalities. Thus the type of culture in a region may have to change, but that should not be a major problem if it occurs slowly enough so adaptive aquaculture management practices can evolve to prevent or minimise the adverse effects of climate change. The presence of aquaculture facilities on coasts can affect local ecosystems, in that the concentrated wastes of cultured organisms can pollute or disturb the environment, sometimes supporting harmful blooms of algae.

Warm water not only holds less oxygen but also accelerates microbial decomposition of aquaculture wastes, lowering oxygen concentrations and perhaps stressing the cultivated organisms as well as organisms near the facility. Higher temperatures can allow for higher incidences of disease, especially if the organisms in the region are under stress. Finally, an increase in the intensity of coastal storms would threaten culture facilities, especially if the region's sea-level rise results in higher storm surges. Overall, the effects of climate change on fisheries and aquaculture will be region-specific, with some regions benefiting and some not.

Influence on Pathogens and Harmful Algal Blooms

Global climate change may affect the abundance and distribution of pathogens and harmful algal blooms, thereby influencing diseases of aquatic organisms as well as human health. For example, the protozoan Perkinsus marinus is the most important pathogen threatening the nation's oyster industry. Low temperatures and salinities usually limit infection by this pathogen and higher temperatures and salinities typically increase the incidence of infection. Higher water temperatures in winter have been linked to the northward spread of Dermo into the northeastern states, so long-term climatic changes may produce shifts in salinity and temperature that favour this oyster pathogen.

Climate change will also affect the distribution of other diseases of aquatic organisms. Warmer coastal waters coupled with eutrophication can also increase the intensity, duration, and extent of harmful algal blooms that damage habitat and shellfish nurseries and that can be toxic to marine species and humans. These blooms appear to be increasing worldwide, perhaps because of increasing eutrophication. The interaction between eutrophication and climate change in possibly promoting harmful algal blooms represents a major area of uncertainty and is a cause for concern because such blooms can have significant effects on marine ecosystems and on human use of these systems. For example, consumption of shellfish that have ingested harmful algae can cause neurotoxic poisoning in humans.

Other Perspectives on Climate Change

There is evidence that marine organisms and ecosystems are resilient to environmental change. Steele hypothesised that the biological components of marine systems are tightly coupled to physical factors, allowing them to respond quickly to rapid environmental change and thus rendering them ecologically adaptable. Some species also have wide genetic variability throughout their range, which may allow for adaptation to climate change. There are at least two schools of thought among physiologists and ecologists who have considered possible effects of climate warming on the survival of marine animals.

Some believe that marine systems experience fewer extinctions of species compared to terrestrial systems because large numbers of marine species have wide geographic temperature ranges as well as greater capacity to migrate to new habitats through their larvae that drift in the water column. Others counter that the seeming lack of evidence for recent marine extinctions is a result of limited information on biodiversity, that not all marine species have drifting larvae, and that many larvae are short-lived or remain relatively close to the parental population before becoming juveniles.

Immediate marine extinctions in response to climate warming are less likely than as a result of direct human disruption of aquatic systems by overexploitation, introduction of exotic species, and coastal pollution. Similarly, the climate warming is unlikely to be a direct cause of extinctions of marine fish and invertebrates and that ecological factors such as interactions with other species may be more significant. Undoubtedly. climate changes will also affect how an organism responds to other environmental stressors.

Global Climate Change on Open Oceans

Influence on Ocean Productivity

Predictions about the effects of climate change on the oceans are often based more on models than observations and are necessarily more general than specific. This is because the immensity of the oceans and the major commitments of shipboard resources necessary for their study have meant that the number of researchers involved and the area covered have been limited. Nevertheless, because the oceans may be affected predominantly by changes in temperature and circulation, predicting the outcome of climate change on the oceans may be somewhat less complicated than for coastal and estuarine systems, where the roles of precipitation and anthropogenic stress are so important.

As noted above for inshore waters, increased temperature or freshwater input to the upper layers of the ocean results in increased density stratification, which affects ocean productivity in two opposing ways. Increased stratification suppresses upwelling of nutrients into the upper, lighted region of the ocean, which leads to decreased production of phytoplankton. At high latitudes, this decrease could be offset somewhat by a lengthened growing season because the mixed layer shallows earlier and deepens later in the year. Where phytoplankton production is limited by the length of the growing season, an increase in this length could increase overall system productivity if suitable nutrients are available.

Coupled physical/biogeochemical models predict a net decrease in global phytoplankton productivity under doubled CO_2 conditions that increase oceanic thermal stratification and reduce upwelling of nutrients. Because productivity varies regionally, simple extrapolation to particular U.S. marine waters is difficult. Although the effects of global climate change cannot be predicted with certainty, they may mimic certain aspects of the effects of natural climatic variability associated with large-scale climate phenomena such as the El Niño-Southern Oscillation (ENSO), the Pacific Decadal Oscillation (PDO), and the North Atlantic Oscillation/Northern Hemisphere Annular Mode (NAO/NAM) that occur on decadal to multi-decadal time scales.

Open ocean productivity is also affected by natural interannual climate variability, and climate-driven changes in the frequency, magnitude, or timing of any of these phenomena may lead to marked changes in water column mixing and stratification, and ultimately a reorganisation of the ecosystem. For example, El Niño-related stratification of the Pacific Ocean around Hawaii over the past decade resulted in a change from a system controlled by nitrogen concentrations to one controlled by phosphorous concentrations; this signals a significant reorganisation of the food webs in the region, with one suite of microorganisms replacing another suite as the dominant primary producers.

Elsewhere, the biological and recruitment effects of ENSO warming and cooling events have been documented for a number of West Coast and Alaska fish species. For example, recruitment increased for Pacific hake, Pacific cod, Pacific herring, and Pacific sardine during strong El Niño events, while West Coast rockfish species suffered recruitment failures. Alaska and West Coast sablefish exhibited opposite growth responses, in that Alaskan sablefish had enhanced growth while West Coast sablefish had retarded growth during El Niño events; this pattern was reversed during La Niña events.

Finally, during the very strong 1982/83 and 1997/98 events, tropical species such as ocean sunfish, Pacific mackerel, and barracuda were found as far north as southeast Alaska. The northward distribution of Pacific mackerel is a particular concern because it is a potential predator of seaward migrating salmon smolts—an example of the potential for food webs to be rearranged by climate change.

Under the PDO, community structure in the Gulf of Alaska ecosystem changed dramatically and abruptly after the climate regime shift of 1976/77 from a cold regime to a warm one. Over a 40-year study period, prey species such as pandalid shrimp and capelin were the dominant species until 1976. After 1977, recruitment of predatory fish increased and, by the 1980s, the prey species had essentially disappeared. Total biomass in the standardised survey catches increased by over 250 percent.

In a broader survey of the biological effects of the 1970s climate shift, Francis et al. documented major changes in the large marine ecosystems of the northeast

Pacific, including abrupt population increases for zooplankton, fish, birds, and marine mammals. The NAO/NAM has also been linked to variations in zooplankton and marine fish production on both sides of the Atlantic. For most commercially important fish species, a positive NAO/NAM results in poor recruitment and growth. The exact mechanism by which these changes occurred is unknown and doubtless differs for many of the species. What appears indisputable, however, is that the driving force behind these widespread ecosystem modifications was climate variability. Anthropogenic climate change can be expected to produce major modifications as well.

Influence on Pacific Salmon

Pacific salmon have been shown to be sensitive to changes in climatic conditions. For example, they have been affected by the natural interannual climate variability described above, so it can be assumed that they will be affected by long-term anthropogenic climate change. Salmon spawn in freshwater, and the young live there for a few months to a few years before spending one to several years in the ocean. Attempts to understand the high variability in the abundances of returning salmon populations have focused on the freshwater phase of their life history, but there is increasing awareness that conditions in the ocean play at least as large a role in modulating population levels and perhaps the dominant role in determining size-at-age of the fish.

There is general agreement that year class strength of salmon is determined during the early rearing period, so conditions in the nearshore environment are critical to survival. These conditions are established by climate-driven and ecological processes, with both local and distant climate processes being important. For example, nearshore production of phytoplankton and zooplankton—

the food of juvenile salmon—reflects not only local atmospheric heating and upwelling, but also the production transported by currents into the region from hundreds of kilometers farther west in the Pacific.

Both ENSO and PDO strongly affect Pacific salmon production from California to northern Alaska. The influence of ENSO is strongest on West Coast populations and diminishes with increasing latitude. The PDO appears to influence both West Coast and Alaska salmon production, although there is an inverse effect on the salmon of the two regions. The most recent positive phase of the PDO from 1977 to 1998 resulted in sustained high Alaska salmon production and a similarly lengthy period of very poor production of West Coast salmon.

During the negative-PDO regime of 1947 to 1976, the relative levels of salmon production between the two regions were reversed; i.e., low in Alaska and high off the West Coast. The climate signatures of ENSO and PDO are fairly similar in the North Pacific, particularly as they affect sea surface temperatures. One effect of global warming may be an increase in the frequency or intensity, or both, of El Niño events. Warmer waters in the North Pacific have coincided with strong Alaska salmon production but with weak West Coast salmon production. However, the amount of warming has been relatively modest; in their positive phases, both the PDO and ENSO increase temperatures by 1 to 3°C.

Whether Alaska salmon would continue to benefit from even higher temperatures is more questionable. Welch et al. showed that salmon habitat is sharply limited by temperature. Under various global warming scenarios and subsequent heating of the North Pacific, they speculate that the habitable region for some salmon species may shrink to encompass only the Bering Sea and not extend south into the Pacific.

Influence on the Thermohaline "Conveyor Belt"

The world's oceans are linked by water movements that are influenced by temperature and salinity. Cold water is denser than warm water, so it sinks below the lighter warm water, and denser high-salinity water sinks below lighter low salinity water. In one instance, this results in a "conveyor belt" system that slowly moves cold dense seawater from the North Atlantic Ocean surface to deep water and then through the Indian and Pacific oceans and back to the Atlantic Ocean over centuries, with the water gradually warming and becoming less dense and moving back toward the ocean surface.

This warm shallow water tempers continental climates as it reaches the eastern Atlantic shores of Europe in the Gulf Stream/North Atlantic current. As the water cools and sinks into the ocean depths in the North Atlantic, it takes with it nutrients, as well as oxygen and CO_2 that it absorbed while at the surface. Deep-sea organisms use the nutrients and oxygen but the CO_2 is not available for use by plants in photosynthesis because of the lack of light in the deep ocean. A weakening or stoppage of this circulation system would diminish vertical mixing between the ocean surface and the deep sea and would reduce upwelling in temperate and subtropical latitudes.

Transport of oxygen from the sea surface to the deep ocean would slow or cease, depriving the deep oceans of their supply of oxygen and slowly rendering them hypoxic or anoxic as happens now in coastal regions. The delivery of nutrients from the upper ocean where plant production occurs to the sunless depths would also be modified, affecting deep-sea animals and their communities in the water column and on the ocean floor. Although the present rate of uptake of CO_2 by the ocean could slow, there is disagreement as to whether or not CO_2 would thereby increase in the atmosphere, with a concomitant increase in global air temperatures, or if the increased stratification of the surface waters would affect biological productivity and thus CO_2 uptake.

There is oceanographic evidence that this thermohaline circulation may have slowed or stopped in the distant past over relatively short time intervals. However, our understanding of the mechanisms involved is rudimentary, hampering predictions about the effects of global climate change in the North Atlantic on the circulation pattern. Coupled general circulation models predict a wide range of responses of the circulation to climate change, from virtually no response to a 40 percent decline sometime in this century. Thus these models must be used cautiously in light of the difficulties with accurately simulating both temperature and salinity.

Changes in the thermohaline circulation are likely with global warming, but that the extent of the changes are uncertain. Based on knowledge to date, Gent and Sun and Bleck do not expect a collapse of thermohaline circulation during the 21st century, due to feedbacks that help stabilise circulation in response to warming. However, they caution that we know little about the feedback mechanisms that affect the circulation pattern. Furthermore, these studies assume atmospheric CO_2 concentrations do not exceed a doubling of pre-industrial levels by 2100.

Earlier modelling efforts using higher baseline greenhouse gas emissions scenarios for the 21st century did simulate the collapse of thermohaline circulation by 2100. Thus, the response of thermohaline circulation to global climate change remains uncertain, with predictions being dependent upon assumptions regarding future greenhouse gas emissions and subsequent temperature increases. There has been little study of the potential ecological consequences of any change in this circulation.

Influence on Sea Ice

One of the more consistently predicted effects of global warming is on high latitude ice cover. Significantly warmer winters in Alaska during the 1980s and 1990s resulted in less sea ice coverage over the Bering Sea. A retreat in the extent of sea ice diminishes the habitat of ringed seals and polar bears that use the ice for hunting, resting, and rearing young. The seals are the main food item for the bears, so changes in their distributions and populations will affect the bears.

Variation in the areal coverage of sea ice and timing of the spring retreat has strong effects on the productivity of the Bering Sea marine ecosystem. For example, the timing of the spring phytoplankton bloom is directly tied to the location of the sea ice edge over the southeast Bering Sea shelf. The presence or absence of a subsurface cold pool, which is better developed and more extensive in summers that follow winters with extensive sea ice coverage, affects the distribution of several commercially important fish stocks, including walleye pollock. Changes in the atmosphere and oceanography of the Bering Sea in the past 20 years have been suggested as driving the tenfold increase in jellyfish biomass over the middle Bering Sea shelf between 1979 and 1997.

Impacts of Climate Change on Coral Reefs

Coral reefs are the most biologically diverse of marine ecosystems, and their economic value includes fishery yields, tourism, and shoreline protection. They occupy 15 to 30 percent of the continental shelf area between 40°N and 40°S. In the United States and its territories, coral reefs occur in Florida, Texas, Puerto Rico, U.S. Virgin Islands, Hawaii, Johnston Atoll, Commonwealth of the Northern Mariana Islands, American Samoa, Guam, and the U.S.-affiliated Pacific Islands. Coral reefs and their associated carbonate environments have been damaged by recent climate changes, so these systems are at risk if climate continues to change.

Coral reef degradation has increased greatly over the last few decades: 27 percent of reefs worldwide are considered to be effectively lost through a combination of anthropogenic effects and extensive coral bleaching episodes. Reefs that have experienced the greatest damage are those near population centres. Examples from the United States include "high-risk" reefs in the U.S. Virgin Islands, Puerto Rico, Johnston Atoll, Tutuilla Island of American Samoa, and some of the Mariana Islands, and "moderate-risk" reefs in the Florida Keys and Hawaii. Reefs in these regions are exposed to a variety of local to regional problems such as pollution from urban and agricultural sources, increased sedimentation due to deforestation, and abusive fishing practices such as the use of dynamite or cyanide. Caribbean corals are also experiencing high mortality from an unexplained increase in a variety of diseases.

Changes in Environmental Factors

In terms of global climate change, coral reefs are affected through a variety of mechanisms that are more or less directly related to increased greenhouse gases in the atmosphere. There are three main consequences of increased anthropogenic greenhouse emissions: seawater chemistry changes, temperature increase, and sea-level rise.

Changes in seawater

The most direct effect of increased CO_2 concentrations on coral reefs involves changes in surface ocean carbonate chemistry. Gas exchange at the air-sea interface allows the partial pressure of CO_2 in the upper ocean to increase proportionately to that in the atmosphere. When carbon dioxide dissolves in water, it forms carbonic acid, a weak acid. An increase in atmospheric CO_2 concentration from preindustrial levels (280 ppmv) to the present level (370 ppmv) has led to the surface ocean becoming more acidic, with its carbonate ion concentration reduced by 15 percent. These shifts in ocean chemistry have been documented by extensive field surveys.

Recent research in experimental systems and in the field with individual species and reef communities indicates that calcium carbonate ($CaCO_3$) precipitation of many reef-building species decreases in response to decreased carbonate ion concentrations. Calculations and model results indicate that under double CO_2 conditions, tropical oceans will experience an average 30 percent decrease in $CaCO_3$ saturation. This will result in an average decrease in coral reef calcification of 20 to 30 percent. The effects of reduced calcification of reef-building corals and calcifying algae are still speculative but will probably be manifested on reefs in the following ways. First, a reduced capacity to secrete $CaCO_3$ requires species to form either more compact or less dense skeletons, or both.

Changes in coral skeletal morphology or strength will thus change a species' functionality or competitiveness in the community, and will most likely lead to a shift in community structure. Second, reduced $CaCO_3$ production in a coral reef community lowers its reef-building capacity. Some reefs may cease to accumulate enough $CaCO_3$ to balance natural erosional processes. Higher latitude reefs such as the northernmost Hawaiian Islands tend to have naturally low carbonate budgets and these reefs are most at risk from reduced reef-building potential. Third, inorganic cementation of reefs is likely to be reduced, leaving them more vulnerable to erosion.

Changes in seawater temperature

The best-known effect of global warming on reefs is coral bleaching, which is a breakdown in the symbiotic relationship between the coral animal and the unicellular algae that live within coral tissues. The algae produce carbohydrates and other photosynthetic products that the coral can use, and are in turn sustained by the coral's waste products that serve as nutrients—a recycling that appears to enhance the productivity of both organisms. This relationship allows corals to thrive in nutrient-poor waters and to secrete massive $CaCO_3$ accumulations. When subject to environmental stress, corals expel the zooxanthellae or the algal pigments or both, leaving the animals pale to colourless.

The symbiotic relationship is thus suspended and the host is weakened until environmental conditions improve and normal zooxanthellae density is regained. Bleaching occurs under a variety of stresses, but the large-scale bleaching events of the last two decades are strongly correlated with increases in maximum monthly sea-surface temperature and exacerbated by increases in solar radiatio. Before the 1980s, coral bleaching events were small-scale phenomena attributed to local stresses such as sudden freshwater influxes or restricted circulation. Regional bleaching events were first noted during the 1982/83 ENSO event when Galápagos reefs suffered severe bleaching and more than 90 percent mortality of coral animals.

Since then, large-scale bleaching events have occurred at all latitudes and in all oceans, and are more pronounced during El Niño years. The worldwide bleaching event in 1998 was the worst ever noted, and destroyed some 16 percent of the world's coral reefs. Nearly all reefs of the United States and its territories have experienced bleaching within the last decade, including significant events in American Samoa, Johnston Atoll, Florida Keys and Puerto Rico. Many corals and reefs can recover from a bleaching event, but the extent of recovery depends on the duration of the event and the severity of the above-normal temperatures.

Barring future episodes of mass mortality, even badly damaged reefs can recover through recolonisation, and it is expected that about half of those destroyed by bleaching in 1997/1998 will recover. Restoration of a coral community usually takes decades and is slowed once reef surfaces are taken over by fleshy algae or other non-reef-building species, which in turn can be exacerbated in those areas where overfishing has removed herbivores. Recovery is also hindered where environmental conditions have become degraded by other stressors such as sedimentation and pollution.

Whether coral reefs will survive continued increases in global warming depends on their ability to adapt physiologically, genetically, or ecologically. For example, the hypothesis that coral bleaching may be an adaptive mechanism by which corals can exchange less heat-tolerant symbionts for more heat-tolerant types has been supported in experiments. Whether this or some similar mechanism will allow corals to adapt to future increases in temperature is uncertain and a topic of debate.

Changes in sea levels

Corals and other reef-building organisms are generally adapted to rising sea levels, and vertical reef growth kept up with a sea-level rise of more than 100 meters since the last glacial period, which included episodic rates exceeding 15 mm/year. Predictions of how well reefs will respond to sea-level rise are mixed because of uncertainties about the magnitude of the rise, local effects of flooding of adjacent lowlands, and complicating effects of other anthropogenic pressures on reefs. Many reefs reached the "growth ceiling" of the sea surface between 6,000 and 3,000 years ago, and have formed expansive reef flats where only the hardiest corals can survive.

On some reefs, flooding these shallow areas will provide extra "head-room" for reef organisms to colonise and grow, albeit temporarily until the reef flats once again accumulate to sea level. On reefs that occur near low-lying coastal areas, sea-level rise is likely to increase coastal erosion rates. This would degrade water quality by reducing light penetration necessary for photosynthesis by zooxanthellae and increasing sedimentation that smothers and stresses coral animals.

Other changes

Cyclones and other tropical storms can occasionally inflict serious damage on coral reefs. If storms increase in frequency or intensity, some reefs will most likely suffer additional damage. If reefs die and are eroded, their role in protecting the adjacent shoreline will be lost, allowing for increased shoreline erosion and silt production during storms. Corals are stressed if large amounts of silt are deposited on them.

Precipitation lowers the salinity of lagoons, which may stress marine organisms, and affects nutrient runoff from land. Increased silt and nutrient runoff may degrade water quality in island lagoons, perhaps causing shifts in the species of resident algae that may lead to harmful algal blooms. Because many species of fish depend on coral reefs for their existence, damaged reefs may no longer support a fishery for small-scale or commercial use.

9

Impacts of Climate Change on Forests

Human activities are the main cause of the climate change. Most of the world's forests have also been extensively modified by human use of the land. Thus, climate and land use are two prongs of human-induced global change. The effect of these forces on forests is mediated by the organisms within forests.

Consideration of climate, land use, and biological diversity is key to understanding forest response to global change. Biological diversity refers to the variety of life at organisational levels from genotypes through biomes. The responses of ecological systems to global change reflect the organisms that are within them.

While ecologists have sometimes not seen the forest for the trees, so to speak, it is also true that forests cannot be understood without knowledge of the trees and other component species. It is the responses of individual organisms that begin the cascade of ecological processes that are manifest as changes in system properties, some of which feed back to influence climate and land use (Figure 1).

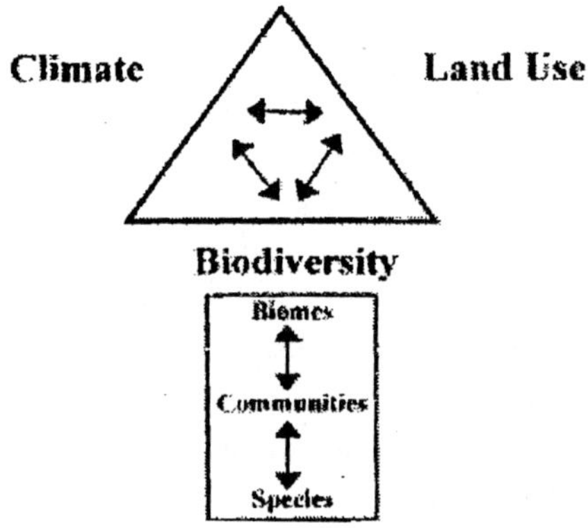

Figure 1. Conceptual model of the aspects of global change.

Beyond its role in ecosystems, biodiversity is invaluable to humans for foods, medicines, genetic information, recreation, and spiritual renewal. Thus, global changes that affect the distribution and abundance of organisms will affect future human well-being and land use, as well as, possibly, the climate.

Influence of Climate Change on Biodiversity

This section focuses on three levels of biodiversity: species, communities, and biomes. Communities are assemblages of interacting species, such as spruce-fir forests or tallgrass prairies. Biomes are major biogeographic regions consisting of distinctive plant life forms. The distributions of species reflect their differential responses to climate, edaphic factors, and biotic interactions, such as competition and herbivory. These species dynamics provide the mechanisms by which communities and biomes respond to global change. At the same time, biomes and communities may constrain the nature of the species' response. For example, species may not be able to shift range without being accompanied by mutualists such as pollinators.

The distribution and abundance of a species are governed by the birth, growth, death, and dispersal rates of individuals comprising a population. These vital rates are, in turn, influenced by environmental factors, including climate, which alter resource availability, fecundity, and survivorship. When aggregated across populations, changes in vital rates manifest as local extinction and colonisation events, which are the mechanisms by which species' ranges change. Some of the best evidence that species' distributions are affected by changing climates is found among plants.

The North American flora shifted during Holocene warming and the rates and direction of response differed among species. The birth, death, and other vital rates of species are also affected by land use. Humans modify the quality, amount, and spatial configuration of habitats. Degradation of habitat quality or quantity can reduce population size and growth rates and elevate the chance of local extinction events. Such habitat loss can reduce genetic diversity, and the ability of species to evolve adaptations to new environments.

Land use may also alter the habitat spatial pattern, increasing the distances among habitat patches. An important consequence of this habitat fragmentation is a reduction in habitat connectivity, which could constrain the ability of many species to move across the landscape in response to climate change. Because of the individualistic responses of species, biotic communities are not expected to respond to climate change as intact units.

Community composition will change in response to a complex set of factors, including the direct effects of climate, differential species dispersal, and indirect effects associated with changes in disturbance regimes, land use, and interspecific interactions. Such an individualistic perspective implies that the impacts of climate change on communities can be understood by the aggregate response across species. However, this perspective may be flawed, because interactions among species are not yet sufficiently understood.

An alternative view suggests that climate change may affect community-level characteristics, such as species richness and resilience to perturbation. Climate may also influence community characteristics by altering the energy available to organisms. Wright et al. found that species richness is often related to ecological productivity, observing frequent correlations with climate, food availability, and limiting nutrients. These factors have all been interpreted to reflect energy availability in a system. Those areas receiving more energy often have a more complex partitioning of usable energy among species, and thus a greater richness, than those areas receiving less energy.

The influence of species richness on ecosystem function is complex. Ecosystems with higher native species richness are sometimes more resilient to perturbation. Also, the presence of exotic species may elevate species richness but inhibit ecosystem function. Nonetheless, the diversity-resilience relationship may be important in anticipating ecosystem response to future climates. Land-use activities also affect the availability of energy in ecosystems. Nearly 40% of the Earth's net primary productivity has been diverted to support human populations. This change in land cover has resulted in the conversion of native habitats supporting diverse species assemblages to intensive land uses that support only simplified, low-diversity communities.

The impacts of eroding biodiversity could include reductions in resilience, resistance to invasion, and ecological services provided to humans. Climate is the primary force shaping the major biomes of the world. Mean and variation in annual precipitation and temperature explain much of the observed pattern of biome distribution. Shifts in biome location depend on the movements of key species. Because the predicted rates of climate change will push the climatic boundaries of biomes northward at a rate faster than the predicted rate of species migration, shifts in biomes will probably lag behind changes in climate.

Land-use activities may appear to be too local in scale to affect regional biome distributions. However, in some biomes, such as the North American Prairie, land conversion is so extensive that little native vegetation remains. Land use can also affect biomes to an extent greater than the sum of the area directly affected. Conversion of natural vegetation to anthropogenic cover types can alter the frequency and scale of disturbance agents that also define the character of biomes. For example, fire suppression and grazing in the American Southwest are thought to be partially responsible for the invasion of woody vegetation into arid grassland habitats.

Climate, Land Use, and Biodiversity

Climate and land use often interact in ways that influence biodiversity, implying that these factors cannot be considered in isolation. For example, land use may modify climatic impacts on species distributions by altering dispersal routes. Where land use creates barriers to dispersal for native species and facilitates dispersal for exotic species, climate change in human dominated landscapes is likely to select for exotic species and against many native species. Such constraints on dispersal are of concern especially around nature reserves.

Organisms "trapped" in reserves by surrounding land use may become extinct if they are not able to disperse to increasingly suitable habitats in other nature reserves. Even in nature reserves, "weedy" species will most likely be quick to replace native species that succumb to climate change. Climate and land use also jointly influence disturbances such as wildfire, flooding, and landslides. Some land uses preset the landscape to be very sensitive to extreme climate events, leading to severe disturbance.

Logging can cause drying of fuels and allow severe fires during normal drought periods. This situation is thought to have transpired in Indonesia, where slash-and-burn agricultural practices provided fuels and ignition sources when an El Niño event induced drought in 1997. Consequently, vast areas of the rainforest burned, possibly jeopardizing many endemic species. Roads and logging practices can similarly increase land sliding and flooding during storm events. Livestock grazing, on the other hand, often reduces fuel loads and reduces wildfire frequency and intensity for a given climate condition.

Changes in land-cover patterns can also directly affect climate, which in turn influences biodiversity. For example, deforestation over large areas can cause reductions in transpiration, cloud formation and rainfall, and increased levels of drying. Such changes can lead to dramatic shifts in biome type, such as the replacement of forests by shrubs or grassland. Similarly, agricultural land use and irrigation in the western Great Plains has been associated with increased cloudiness and precipitation in the Rocky Mountains of Colorado.

Trends in biodiversity to changes in climate and land use over the last 100 years provide a context for understanding future interactions. During the past century, average air temperatures have increased 0-2 °C over most of the United States and Canada. Changes in precipitation have been variable over this period, increasing from 5% to 20% over most of the United States, but decreasing up to 20% in the Northern Rockies and California. The US population grew from 76 million in 1900 to over 270 million in 1998. Agricultural lands increased in area until 1900 then decreased slowly until 1950; they have remained stable since then.

Between 1942 and 1992, urban area increased 120%, while protected areas increased 80%. Rural residential development has increased rapidly in the mountain west in recent decades, expanding human influences into semi-natural habitats. Many species and communities have responded to these changes in climate and land use. For example, forest decline and dieback are evident along the Atlantic and Pacific coasts and may be related to elevated CO_2 levels and climate change. The breeding ranges of some mobile species, such as waterfowl, have been expanding northward in association with climate amelioration.

Shifts in demography are apparent in some species. Both amphibians and birds in Great Britain have shifted breeding dates by 1 to 3 weeks earlier since the 1970s in association with increasing temperatures. Species associated with human-dominated landscapes have greatly expanded in recent years. These include many large ungulate and small mammal species, waterfowl, and some bird species associated with agricultural and urban environments. Some of these species are now so abundant that the primary concern is controlling their populations. Many exotic species have also become established in the United States and greatly expanded their ranges.

At the same time, several natural community types and numerous species have been greatly reduced by human activities. For example, natural spruce-fir, longleaf pine, and loblolly-shortleaf pine forests now cover less than 2% of their presettlement ranges and are likely to be further reduced under global warming. Many species dependent upon these endangered and reduced ecosystems are currently in peril.

The number of species listed as threatened and endangered in the United States under the Endangered Species Act currently totals 1232. Factors contributing to species endangerment include habitat conversion, resource extraction, and exotic species. The spatial distribution of such factors results in species at risk being concentrated in particular regions of the United States, especially the southern Appalachians, the arid Southwest, and coastal areas.

Future Challenges

Given past relationships among climate, land use, and biodiversity, how might biodiversity respond to future global change? Some researchers assess potential vegetation response to projected future climate change for doubled CO_2 concentrations by using the climate predictions of general circulation models (GCMs) as input to a set of different vegetation simulation models. The primary focus is on the potential continental-scale response of forest vegetation as reflected in changes in the distributions of biomes, community types, and tree species. Although vegetation models incorporating landuse dynamics are under rapid development at local scales, the current state of knowledge does not allow for integrating the effects of land use at the continental scale.

In the assessment, they chose to emphasise forests as a key element of biodiversity, which allowed us to draw implications for other organisms that require forest habitats. They also simulate changes in species richness of trees and terrestrial vertebrates based on energy theory and examine potential effects of climate change on locations where endangered species are concentrated. Results are for the coterminous United States, unless stated otherwise.

The models used to anticipate biodiversity responses to changes in climate are based on empirical relationships between organisms and the environments they currently occupy. By using these relationships to predict biodiversity response, the researchers are implicitly assuming that these environment-organism relationships will remain unaltered in the future. Although this approach can be criticised for not modelling some or all of the actual mechanisms leading to shifts in vegetation and species ranges, such an approach is commonly taken to initially assess ecological responses to global change scenarios. It is also important to emphasise that the given GCMs are coarsegrid, regionally smoothed outputs that do not allow depiction of local or even subregional climates. Consequently, the vegetation outputs modelled here must also be considered coarse and not sensitive to local phenomena.

Climate change scenarios: The GCMs differ in formulation, hence predictions vary among the models. Consequently, the simulations are best considered as possible alternative views of the future, with unknown likelihood of occurrence. Both equilibrium and transient GCM scenarios are used in this assessment to incorporate a broad array of possible futures. Still, the uncertainty level for each climate and vegetation output is unknown, and probably high, given the uncertainty in forecasting climate and subsequent vegetation responses.

For the coterminous United States, the different climate scenarios all predict some level of warming and increased annual precipitation. Mean annual temperature increases vary from 3.3°C to 5.8°C, with the greatest warming at higher latitudes. Mean annual precipitation is predicted to increase in the West, with 20% to more than 50% increases in California. Decreased precipitation of up to 30% is predicted for locations in the Southeast, Texas, and the Northwest.

Biomes: The MAPSS biogeography model projects biome response to climate as change in vegetation structure and density based on light, water, and nutrient limitations. Vegetation is coupled directly to climate and hydrology, and rules are applied to classify vegetation into biome types. The model considers the effects of altered CO_2 on plant physiology. MAPSS simulates potential natural vegetation based on climate and does not include succession or plant dispersal. The results project that potential forest area decreases by an average of 11% across the GCM scenarios, with a range of +23% under the coolest scenarios and -45% under the hottest scenarios.

Northeast mixed forests decrease by 72% in potential area, on average, as they shift into Canada and increase in potential area continentally. The potential area of eastern hardwoods decreases by an average of 34%. These deciduous forests shift north, replacing northeastern mixed forests, but are squeezed from the south by southeastern mixed forests or from the west by savannas and grasslands, depending on the scenario. The potential range of southeastern mixed forests increases under all scenarios while shifting north. This biome remains intact under cooler scenarios but is converted to savannas and grasslands in the South under the hotter scenarios.

Potential environments for alpine ecosystems all but disappear from the western mountains, being overtaken by encroaching forests. Wet coniferous forests in the

Northwest decrease in potential area by 9% on average. The potential range of interior western pines change little on average, with a range of -30% to +39%. The potential range of shrublands and arid woodlands expand in the interior West and Great Plains, encroaching on some grasslands. However, grassland habitats may expand in the deserts of the Southwest, parts of the Southeast, and possibly in the upper Midwest. Thus, the potential area of grassland could either decrease or increase.

The projections agree on a single or on two vegetation classes across 68% of the coterminous United States. Locations of greatest certainty are in the northern plains and Florida. Regions of great uncertainty are transition zones in the eastern prairie and the West. Taken in order of increasing temperature change, the future scenarios imply that potential forest range could expand with small amounts of warming but would contract under the hotter scenarios.

Forest community types and tree species: Statistical models are used to project the distributions of tree species, with results aggregated into community types. These models project tree response to future climate, based on current relationships between trees and environmental variables such as climate and soils. Because physiological data are not required, many species can be modelled. However, these approaches do not include important forest dynamics involving species interactions, physiological processes such as CO_2 uptake, and tree dispersal and establishment. They also assume that species-environment relationships will remain the same under future climate conditions, which may not be the case.

Two statistical models are used. The DISTRIB model of Iverson and Prasad was applied to the eastern United States. This model uses regression tree analysis, based on

33 environmental variables, to predict the potential future distribution of suitable habitat for 80 tree species. An index of species response was derived by multiplying the importance value by the area for each county. Community types were then defined by simply aggregating the importance values for individual tree species.

A local regression model is used in western forests to predict probability of occurrence of tree species across North America based on climate and soils. The results for dominant western species are presented. In the eastern United States, the area of habitat suitable for oak-hickory expands in area following climate change by an average of 34%, primarily to the north and east. The oak-pine habitat also expands by roughly 290% and is represented throughout the Southeast. On the other hand, the habitat capable of supporting the spruce-fir and aspen-birch types are dramatically reduced and are largely replaced by oak-hickory and oak-pine habitat.

The loblolly-shortleaf pine habitat is also reduced by 32%; it shifts to the north and west while being replaced in its current zone by oak-pine habitat. The longleaf-slash pine habitat is reduced, on average, by 31%. There is a higher level of disagreement among climate scenarios for the elm-ash-cottonwood, oak-gum-cypress, and white-red-jack pine types, which show increases in habitat under some scenarios and decreases under others. These potential changes in habitat for community types reflect the responses of individual tree species.

Seven of the 80 species modelled were predicted to have their suitable habitat reduced in regional importance by at least 90%: bigtooth aspen (Populus grandidentata), quaking aspen (P. tremuloides), sugar maple (Acer saccharum), northern white cedar (Thuja occidentalis), balsam fir (Abies balsamea), red pine (Pinus resinosa), and paper birch (Betula papyrifera). An additional 24 species would potentially decline by at least 10%. In contrast, regional importance increased for 35 species, including 12 species that increased suitable habitat by 100% or more, including four species of oak and one hickory. Most species' habitat was projected to move to the north, 100 to 530 km for several species.

Some species, such as quaking aspen, paper birch, northern white cedar, balsam fir, and sugar maple have the optimum latitude of suitable habitat move north of the US border. All of the above analyses for the eastern United States relate to the potential distribution of suitable habitat, not actual distribution of the species. The assumption is that the species will get there, that there are no barriers or constraints to migration. In the western United States, the potential ranges of dominant rainforest conifers such as western hemlock are projected to decrease west of the Cascade Mountains and expand into mountain ranges throughout the interior West.

Simulated potential habitat for Douglas fir also decreases along the western coast of the coterminous United States but expands east of the Cascades and Sierras as well as northward along the west coast of Canada into Alaska. The potential range of several subalpine conifers is simulated to contract substantially in the western coterminous United States, including Englemann spruce, mountain hemlock, and several species of true fir. Potential ranges for these subalpine species are simulated to expand along the western coast of Canada and into Alaska. Potential future habitat for big sagebrush, an important shrub in the inland West, is largely absent in the United States, shifting into Canada. This shrub is simulated to be replaced in the United States by shrubs, chaparral, and grasslands, now found in arid regions of the Southwest.

Potential habitat for ponderosa pine is simulated to expand in the western United States, including on the West Coast, where many other conifers are projected to contract. While the potential ranges of many taxa in the West shift northward, the topographic complexity of the region creates less intuitive changes in some species distributions. Potential habitat for some conifer species associated with mesic climates shifts south and east along the Rocky Mountains with, for example, forests typical today of Glacier National Park becoming dominant in Yellowstone National Park to the southeast.

The direction of change may also differ from east to west within a species' range. Potential habitat for paper birch, whose range spans the continent, is simulated to contract northward in the eastern United States but to expand southward along the Rocky Mountains in the West. The complex topography of the West results in simulated future habitat for many tree species that is disjunct. Consequently, dispersal to new habitats under climate change may be more difficult in the West than in the East, where the future distributions of species are simulated to be more continuous.

Community richness: Currie derived multiple regression models relating the broad-scale variability of species richness of trees and terrestrial vertebrates across North America to the spatial patterns of summer and winter precipitation and temperature. These models were used to predict patterns of species richness under five GCM scenarios. This approach assumes that richness will continue to covary with climate in the future in the same way that it does today. The results indicate that contemporary patterns of richness correlate strongly with temperature and less strongly with precipitation.

These climate-richness relationships differ among taxonomic groups; thus, projected change in richness under climate change also differed among groups. Current tree species richness is a positive function of temperature up to relatively high temperatures, then is negatively related to further increases in temperature. Current tree species richness is also a positive function of precipitation. Thus, climatic warming is predicted to lead to increased tree richness over most of the northern United States, especially in the areas that are now coldest: in the western mountains and near the Canadian Border.

Moderate decreases in richness (-20%) are predicted to occur in areas that are likely to experience drying and very high temperatures, such as the southwestern deserts. The greatest disagreement among the projections was for Pacific Northwest. All models predict increases in richness in that area, but these increases may be modest (< +50%) to pronounced (> +100%). Contemporary species richness of endotherms covaries strongly with temperature. Richness in these groups is maximal in moderately warm areas, and it decreases in hotter areas. This relationship may occur because ambient heat serves as a direct energy subsidy for endotherms in cold climates, but these organisms expend energy to dissipate heat in hot areas.

Endotherm richness is only weakly related to precipitation. Consequently, under most climate change scenarios, endotherm richness is predicted to decrease by over 25% in low-elevation areas in the Southeast. Increases in richness (> +11% to > +100%), are predicted for upper montane areas across the United States. Contemporary ectotherm richness is even more strongly related to temperature, increasing monotonically as temperature increases. Ambient heat is also an energy subsidy for ectotherms, even in the warmest areas of the earth. Consequently, climatic warming is predicted to increase ectotherm richness over the entire coterminous United States.

For reptiles, the increase is predicted to be modest across the South and greater in the North. Amphibian richness, in contrast, also depends somewhat on precipitation. Because most GCMs predict that the southeastern United States will become somewhat drier in winter, amphibian richness in the Southeast is predicted to change little despite increased temperatures, and to increase elsewhere. Differences among predictions of different GCMs lie mainly in how dramatically richness is likely to increase.

Implications of Assessment Results

In contrast to public perceptions of large-scale forest loss under global change, the biome models project only a modest average loss of forest area (11%) for the coterminous United States. Lost forest is largely replaced by savanna and arid woodland biome types. However, the projected response of forest habitats to global change scenarios is highly variable among GCMs. Forest habitats are projected to increase by 23% under the HADCM2SUL model and are predicted to decrease by 45% under the UKMO model. This uncertainty points to cautious interpretation of the findings and to the need for further research directed at understanding the factors driving change in forest systems.

Within the climate projections considered, the researchers did observe some common patterns in the response of forest community types across the continent. Areal expansion of habitats was projected for oak-hickory and oak-pine in the East and ponderosa pine and arid-tolerant hardwoods in the West. The first three of these are extremely valuable for forest products and as habitat. The heavy mast production of oak-hickory, for example, is a food source for more than 180 different kinds of vertebrates. The oak types also produce persistent coarse woody debris that benefits several ecological processes and organisms.

Suitable habitats for several important community types, however, are projected to greatly decrease in area or disappear from the coterminous United States. These include alpine habitats, subalpine spruce-fir forests, aspen, the maple-beech-birch type, sagebrush, and loblolly-shortleaf pine communities. Subalpine spruce-fir has been decreasing in area, becoming increasingly fragmented, and losing species richness since the last glacial period. Other habitats such as sagebrush and aspen are being reduced in modern times by conifer encroachment, grazing, and exurban development.

Climate warming is projected to further reduce these community types and the multitude of species dependent upon them. Quaking aspen, for example, is a keystone species in the west, where it is often the only abundant hardwood among the conifer forests. Its palatability, soft wood, and rapid decomposition result in many species of plants and animals being dependent upon it. Subalpine spruce-fir now supports several species that require wilderness conditions. Sagebrush has several obligate species of vertebrates.

The maple-beech-birch type includes several important tree species, such as red maple, sugar maple, black cherry, American beech, and yellow birch. Reductions of these species will influence many associated organisms. Suitable habitats for the forest types projected to decrease in the coterminous United States are generally expected to increase in Canada. Some of these northerly locations are underlain by permafrost and have nutrient-poor soils.

More work is needed to determine which species could tolerate future conditions in the northern ecosystems.

Important policy questions arise from the cross-border migration of species and ecosystems that are increasingly threatened in the United States but increasingly common in Canada. Such changes might suggest that the current national regulations and incentives on biodiversity be supplemented with international regulations and incentives.

Climate change will favour greater tree species richness over much of the coterminous United States. Climatic conditions are also projected to become more favourable for amphibians and reptiles. Whether these climate changes will counter the current decline in amphibians caused by pollution, ultraviolet radiation, land use, and other factors will require further study. Climate conditions are predicted to lead to lower bird and mammal richness across the southern United States. What are the implications of this projection for the many bird species that winter in the southern United States and breed to the north?

Perhaps of greatest concern are currently threatened or endangered bird and mammal species that are restricted to endangerment hotspots in the Southeast. The projected losses of species richness here could further imperil these populations. Individual species studies will be needed to begin to understand the implications of these changes. These projections for species richness raise interesting questions for management. If climate change increases the potential for amphibian and reptile species richness, which species are likely to disperse to the newly suitable locations? (Fig. 2) Are there opportunities to introduce desirable species to the newly suitable habitats and select against nondesirable species that are good dispersers? The projections for birds and mammals indicate no change or slight increases in richness in the North, but decreases in the South. Which species in the southern areas are most likely to go extinct? Can management strategies be used to buffer these losses?

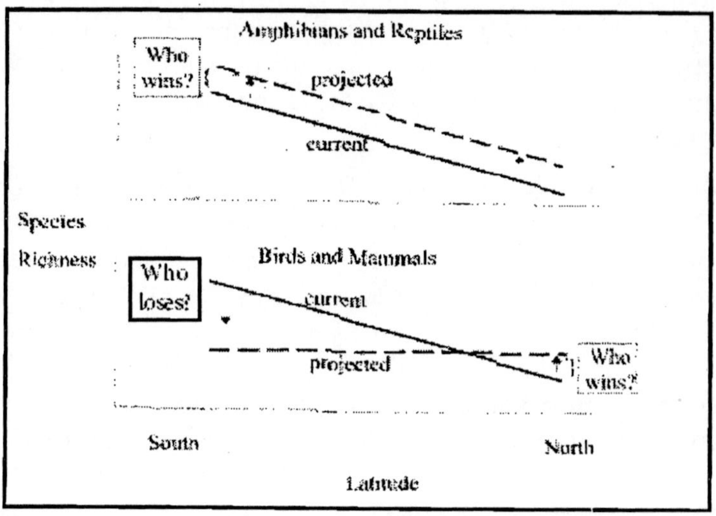

Figure 2. Current trends in species richness along a latitudinal gradient and projected trends under climate change

Local variation in climate due to topography or other factors could result in species or communities being able to persist in suitable microhabitats, even though the coarser-resolution models project no suitable habitats in the location. The current locations of species, communities, and biomes reflect environmental conditions suitable for all life-history stages of these organisms. In reality, however, the habitat requirements for seedling establishment may not be identical to those for the survival and reproduction of adult trees and shrubs. The likelihood that organisms will be able to disperse to newly suitable habitats will vary considerably among species. Thus, differential rates of species dispersal will be a key determinant of future biodiversity patterns.

During the Holocene, dominant plant species migrated at rates that allowed them to keep pace with climate change. However, the estimated rates of dispersal during the Holocene, about 10-45 km per century, were much slower than the potential geographic shifts implied by the analysis of forest communities. It is unclear how fast organisms may disperse in modern landscapes subjected to various levels and types of land uses. It is likely that dispersal rates will be slower than the rate of climate change and that weedy species will be better able than many other species to disperse through human-dominated landscapes.

These differential rates of dispersal suggest that actual plant communities under climate change will not resemble those predicted based on community-environment relationships. Instead, the new communities will initially be dominated by the subset of species that are best able to track climate change. Complex vegetation dynamics should then follow as better competitors gradually arrive at these sites. Adding to the complexity are the dynamics of the species that occupied the site under previous climatic conditions.

While some species are likely to vacate unsuitable sites rather quickly, others may persist for decades to centuries. Franklin et al. suggested that adult trees in old-growth forests in the Pacific Northwest may persist long after changing climate no longer allows regeneration. These forests could then change quickly as senescence or disturbance clears the site, allowing emigrating species to establish. A final caveat on the projections is that they do not consider land use, disturbance, or fine-scaled vegetation dynamics.

Some community types, such as prairie and coastal chaparral, are now dominated by intense human land use, which has altered the expected pattern of disturbance and negatively influenced native species. Of the 63 factors that have contributed to species endangerment in the United States, land-use intensification associated with residential and urban development, forest management, grazing, and environmental contamination are the most common factors cited in nine out of the 12 endangerment hotspots.

As the population of the United States increases, the area of seminatural habitats will be further reduced, causing deviations from the relationship between climate and biodiversity. This list of limitations makes evident the complexity of the interaction of global change and biodiversity. Although it is important for researchers to attempt to predict future global change and consequences for biodiversity, society needs to be aware that the accuracy of these predictions may be mixed. The complex suite of interactions initiated by climate change could well lead to some unexpected outcomes. How to cope with such surprises may be one of the greatest challenges of future global change.

Coping Strategies

Managing global climate-change impacts on biodiversity involves avoidance of impending climate and land-use changes, altering those changes, or accepting the changes and dealing with their impacts. Strategies to slow global change include reducing the atmospheric concentration of greenhouse gases, human-induced disturbances, and land-cover changes. Control of greenhouse gas emissions requires reduced use of fossil fuels and less harvesting of large trees for short-turnover products, as well as the establishment of new locations or means of carbon storage by such actions as planting large areas with rapidly growing trees or enhancing their carbon sequestration potential.

However, carbon storage in large plantations of monocultures of nonnative species would jeopardize native diversity if those species replace natural vegetation. Strategies for reducing changes in land cover and use include management of human population growth, land-use planning, and land-use regulation and incentives programmes. These strategies can be designed to foster biodiversity, if that goal is included in the overall plan. Where the maintenance of ecosystem processes and native species is a priority, effective strategies may differ with location, community type, and management objective.

For communities that are unlikely to migrate to suitable environments elsewhere, it may be appropriate to minimise change by manipulating vegetation structure, composition, or disturbance regimes to favour the current community. For communities that may be able to reach newly suitable habitats, a reasonable strategy may be to manage some of the current habitat as a reservoir until the community is reestablished in the new locations. Other portions of the current habitat may be managed to encourage change to the new species and communities more appropriate for the new environment.

Global change could offer opportunities to restore communities that are now degraded. In this case, management to induce rapid change may allow for the establishment of species deemed to be desirable by society. In some cases diversity can be preserved only through reserves set aside to protect species in the face of global land changes. Halpin offers a strategic framework of this type for nature reserves. The framework involves both maintaining current communities and facilitating natural dispersal of organisms across elevational and latitudinal gradients. The five categories of management prescriptions presented by Halpin are

1. selection of redundant reserves,
2. selection of reserves that provide habitat diversity,
3. management for buffer-zone flexibility,
4. management for landscape connectivity, and
5. management for habitat maintenance.

The exact prescription will vary, depending on characteristics of the species and communities to be preserved and their habitats. For the species most at risk, seed banks and captive breeding and rearing approaches may be necessary until new suitable habitats develop. Much work is needed on methods to execute and evaluate such strategies. Simulation models and other decisiònsupport tools are needed to assess the likely response of a community to global change and to evaluate the potential success of alternative management strategies.

It is especially important that these models integrate consideration of climate, land use, and biodiversity. Adequate monitoring protocols are needed to establish rates of change in environmental drivers and species and community responses to these changes. Adaptive management experiments can be used to evaluate the success of management manipulations. These experiments should include manipulations of species distributions and performance through planting and release projects, habitat modification, genetic engineering, and eradication of undesirable species. In addition, species banks or refuges for colonisation can be developed.

Decisions on mitigation of the effects of climate change on biodiversity involve social, political, and economic considerations. Therefore, these decisions will not be made solely within the ecological context of the issues. Instead, the full set of advantages and disadvantages of all decisions must be considered.

However, choosing not to control greenhouse gases or not to manage species, communities, and landscapes in the face of climate change is making a decision about the impact of these changes on biodiversity. Rather than letting inaction decide the result of potential changes, active assessment and management will be necessary to bring the landscape to a state of desired future biodiversity conditions in the face of global climate change.

IMPORTANCE OF RESEARCH ON GLOBAL CLIMATE CHANGE

Considerable progress has been made toward understanding global change, major research initiatives will be required to reduce current uncertainty.

Future climate and land use: The predictions of current GCMs disagree to varying degrees and their levels of accuracy are not well quantified. The models could be better validated relative to past conditions at local to regional scales. This would allow research and management to focus on the future climate scenarios that are most plausible. In contrast to climate, relatively little effort has gone into understanding and predicting land-use and land-cover change. Studies are needed to project land cover and use based on biophysical factors and socioeconomic factors. Moreover, integrated models of climate and land use are needed to better predict future interactions between these two aspects of global change.

Tolerances of organisms: Much is known about the tolerances of some trees to climate, soils, and other biophysical controls. However, such knowledge is mostly lacking for other plants, vertebrates, and invertebrates. Similarly, little is known about the demographic performance of organisms across gradients in climate and land use. This deficiency is especially apparent for population dispersal. Species at higher trophic levels will probably be more difficult to model under global change because they respond directly to climate as well as indirectly to the secondary effects mediated by habitat structure, ecological productivity, and interactions with other species. Thus, holistic examinations of species and environmental relationships are needed that consider multiple stressors and multiple spatial and temporal scales.

Biodiversity feedbacks: Organisms mediate the effects of global change on ecosystems and feedbacks to climate and land use. What might be the consequences on the services provided by ecosystems of the changes in biodiversity predicted under global change? Knowledge of the role of biodiversity in ecosystem function is underdeveloped. More research is needed on how species composition feeds back to influence climate and land use.

10

Mitigation of Climate Change

Climate change is a problem with unique characteristics. It is global, long-term, and involves complex interactions between climatic, environmental, economic, political, institutional, social and technological processes. This may have significant international and intergenerational implications in the context of broader societal goals such as equity and sustainable development. Developing a response to climate change is characterised by decision-making under uncertainty and risk, including the possibility of non-linear and/or irreversible changes.

Alternative development paths can result in very different greenhouse gas emissions. Development paths leading to low emissions depend on a wide range of policy choices and require major policy changes in areas other than climate change. Climate change mitigation will both be affected by, and have impacts on, broader socio-economic policies and trends, such as those relating to development, sustainability and equity.

Climate mitigation policies may promote sustainable development when they are consistent with such broader societal objectives. Some mitigation actions may yield extensive benefits in areas outside of climate change: for example, they may reduce health problems; increase employment; reduce negative environmental impacts; protect and enhance forests, soils and watersheds; reduce those subsidies and taxes which enhance greenhouse gas emissions; and induce technological change and diffusion, contributing to wider goals of sustainable development.

Similarly, development paths that meet sustainable development objectives may result in lower levels of greenhouse gas emissions. Differences in the distribution of technological, natural and financial resources among and within nations and regions, and between generations, as well as differences in mitigation costs, are often key considerations in the analysis of climate change mitigation options. Much of the debate about the future differentiation of contributions of countries to mitigation and related equity issues also considers these circumstances.

The challenge of addressing climate change raises an important issue of equity, namely the extent to which the impacts of climate change or mitigation policies create or exacerbate inequities both within and across nations and regions. Developed countries and countries with economies in transition limit and reduce their greenhouse gas emissions first.

Techniques for Reducing Mitigation

Significant technical progress relevant to greenhouse gas emissions reduction has been made since the SAR in 1995 and has been faster than anticipated. Advances are taking place in a wide range of technologies at different stages of development, e.g., the market introduction of wind turbines, the rapid elimination of industrial by-product gases such as N_2O from adipic acid production and perfluorocarbons from aluminium production, efficient hybrid engine cars, the advancement of fuel cell technology, and the demonstration of underground carbon dioxide storage.

Technological options for emissions reduction include improved efficiency of end use devices and energy conversion technologies, shift to low-carbon and renewable biomass fuels, zero-emissions technologies, improved energy management, reduction of industrial by-product and process gas emissions, and carbon removal and storage. Following are some key findings from many studies:

— Hundreds of technologies and practices for end-use energy efficiency in buildings, transport and manufacturing industries account for more than half of this potential.

— At least up to 2020, energy supply and conversion will remain dominated by relatively cheap and abundant fossil fuels. Natural gas, where transmission is economically feasible, will play an important role in emission reduction together with conversion efficiency improvement, and greater use of combined cycle and/ or co-generation plants.

— Low-carbon energy supply systems can make an important contribution through biomass from forestry and agricultural by-products, municipal and industrial waste to energy, dedicated biomass plantations, where suitable land and water are available, landfill methane, wind energy and hydropower, and through the use and lifetime extension of nuclear power plants. After 2010, emissions from fossil and/or biomass-fuelled power plants could be reduced substantially through pre- or post-combustion carbon removal and storage. Environmental, safety, reliability and proliferation concerns may constrain the use of some of these technologies.

— In agriculture, methane and nitrous oxide emissions can be reduced, such as those from livestock enteric fermentation, rice paddies, nitrogen fertilizer use and animal wastes.

— Depending on application, emissions of fluorinated gases can be minimised through process changes, improved recovery, recycling and containment, or avoided through the use of alternative compounds and technologies.

Half of these potential emissions reductions may be achieved by 2020 with direct benefits exceeding direct costs, and the other half at a net direct cost of up to US$100/tCeq. These cost estimates are derived using discount rates in the range of 5% to 12%, consistent with public sector discount rates. Private internal rates of return vary greatly, and are often significantly higher, affecting the rate of adoption of these technologies by private entities. Depending on the emissions scenario this could allow global emissions to be reduced below 2000 levels in 2010-2020 at these net direct costs.

Realising these reductions involve additional implementation costs, which in some cases may be substantial, the possible need for supporting policies, increased research and development, effective technology transfer and overcoming other barriers. Studies do not exist for every sector and region. Forests, agricultural lands, and other terrestrial ecosystems offer significant carbon mitigation potential. Although not necessarily permanent, conservation and sequestration of carbon may allow time for other options to be further developed and implemented. Biological mitigation can occur by three strategies:

(a) conservation of existing carbon pools,

(b) sequestration by increasing the size of carbon pools, and

(c) substitution of sustainably produced biological products, e.g. wood for energy intensive construction products and biomass for fossil fuels.

Conservation of threatened carbon pools may help to avoid emissions, if leakage can be prevented, and can only become sustainable if the socio-economic drivers for deforestation and other losses of carbon pools can be addressed. Sequestration reflects the biological dynamics of growth, often starting slowly, passing through a maximum, and then declining over decades to centuries. Conservation and sequestration result in higher carbon stocks, but can lead to higher future carbon emissions if these ecosystems are severely disturbed by either natural or direct/indirect human-induced disturbances.

Even though natural disturbances are normally followed by re-sequestration, activities to manage such disturbances can play an important role in limiting carbon emissions. Substitution benefits can, in principle, continue indefinitely. Appropriate management of land for crop, timber and sustainable bio-energy production, may increase benefits for climate change mitigation. Taking into account competition for land use and the SAR and SRLULUCF assessments, the estimated global potential of biological mitigation options is in the order of 100GtC (cumulative), although there are substantial uncertainties associated with this estimate, by 2050, equivalent to about 10% to 20% of potential fossil fuel emissions during that period.

Realisation of this potential depends upon land and water availability as well as the rates of adoption of different land management practices. The largest biological potential for atmospheric carbon mitigation is in subtropical and tropical regions. Cost estimates reported to date of biological mitigation vary significantly from US$0.1/tC to about US$20/tC in several tropical countries and from US$20/tC to US$100/tC in non-tropical countries. Methods of financial analysis and carbon accounting have not been comparable.

Moreover, the cost calculations do not cover, in many instances, inter alia, costs for infrastructure, appropriate discounting, monitoring, data collection and implementation costs, opportunity costs of land and maintenance, or other recurring costs, which are often excluded or overlooked. The lower end of the ranges are biased downwards, but understanding and treatment of costs is improving over time. These biological mitigation options may have social, economic and environmental benefits beyond reductions in atmospheric CO_2, if implemented appropriately. However, if implemented inappropriately, they may pose risks of negative impacts.

Biological mitigation options may reduce or increase non-CO_2 greenhouse gas emissions. There is no single path to a low emission future and countries and regions will have to choose their own path. Most model results indicate that known technological options could achieve a broad range of atmospheric CO_2 stabilisation levels, such as 550ppmv, 450ppmv or below over the next 100 years or more, but implementation would require associated socio-economic and institutional changes. To achieve stabilisation at these levels, the scenarios suggest that a very significant reduction in world carbon emissions per unit of GDP from 1990 levels will be necessary.

For the crucial energy sector, almost all greenhouse gas mitigation and concentration stabilisation scenarios are characterised by the introduction of efficient technologies for both energy use and supply, and of low- or no-carbon energy. However, no single technology option will provide all of the emissions reductions needed. Reduction options in non-energy sources and non-CO_2 greenhouse gases will also provide significant potential for reducing emissions.

Transfer of technologies between countries and regions will widen the choice of options at the regional level and economies of scale and learning will lower the costs of their adoption. Social learning and innovation, and changes in institutional structure could contribute to climate change mitigation. Changes in collective rules and individual behaviours may have significant effects on greenhouse gas emissions, but take place within a complex institutional, regulatory and legal setting. Several studies suggest that current incentive systems can encourage resource intensive production and consumption patterns that increase greenhouse gas emissions in all sectors, e.g. transport and housing.

In the shorter term, there are opportunities to influence through social innovations individual and organisational behaviours. In the longer term such innovations, in combination with technological change, may further enhance socio-economic potential, particularly if preferences and cultural norms shift towards lower emitting and sustainable behaviours. These innovations frequently meet with resistance, which may be addressed by encouraging greater public participation in the decision-making processes. This can help contribute to new approaches to sustainability and equity.

Benefits of Mitigation

Estimates of cost and benefits of mitigation actions differ because of

1. how welfare is measured,
2. the scope and methodology of the analysis, and
3. the underlying assumptions built into the analysis.

As a result, estimated costs and benefits may not reflect the actual costs and benefits of implementing mitigation actions. With respect to 1 and 2, costs and benefits estimates, inter alia, depend on revenue recycling, and whether and how the following are considered: implementation and transaction cost, distributional impacts, multiple gases, land-use change options, benefits of avoided climate change, ancillary benefits, no regrets opportunities and valuation of externalities and non-market impacts. Assumptions include, inter alia:

- Demographic change, the rate and structure of economic growth; increases in personal mobility, technological innovation such as improvements in energy efficiency and the availability of low-cost energy sources, flexibility of capital investments and labour markets, prices, fiscal distortions in the no-policy (baseline) scenario.
- The level and timing of the mitigation target.
- Assumptions regarding implementation measures, e.g. the extent of emissions trading, the Clean Development Mechanism (CDM) and Joint Implementation (JI), regulation, and voluntary agreements and the associated transaction costs.
- Discount rates: the long time scales make discounting assumptions critical and there is still no consensus on appropriate long-term rates. These discount rates should be distinguished from the higher rates that private agents generally use in market transactions.

Some sources of greenhouse gas emissions can be limited at no or negative net social cost to the extent that policies can exploit no regrets opportunities:

- *Market imperfections:* Reduction of existing market or institutional failures and other barriers that impede adoption of cost-effective emission reduction measures, can lower private costs compared to current practice. This can also reduce private costs overall.

— *Ancillary benefits:* Climate change mitigation measures will have effects on other societal issues. For example, reducing carbon emissions in many cases will result in the simultaneous reduction in local and regional air pollution. It is likely that mitigation strategies will also affect transportation, agriculture, landuse practices and waste management and will have an impact on other issues of social concern, such as employment, and energy security. However, not all of the effects will be positive; careful policy selection and design can better ensure positive effects and minimise negative impacts. In some cases, the magnitude of ancillary benefits of mitigation may be comparable to the costs of the mitigating measures, adding to the no regrets potential, although estimates are difficult to make and vary widely.

— *Double dividend:* Instruments provide revenues to the government. If used to finance reductions in existing distortionary taxes, these revenues reduce the economic cost of achieving greenhouse gas reductions. The magnitude of this offset depends on the existing tax structure, type of tax cuts, labour market conditions, and method of recycling. Under some circumstances, it is possible that the economic benefits may exceed the costs of mitigation.

The cost estimates for Annex B countries (Group of countries included in Annex B in the Kyoto Protocol that have agreed to a target for their greenhouse gas emissions, including all the Annex I countries but Turkey and Belarus.) to implement the Kyoto Protocol vary between studies and regions as indicated in Paragraph 11, and depend strongly upon the assumptions regarding the use of the Kyoto mechanisms, and their interactions with domestic measures. The great majority of global studies reporting and comparing these costs use international energy-economic models. Nine of these studies suggest the following GDP impacts :

Annex II countries (Group of countries included in Annex II to the UNFCCC, including all developed countries in the Organisation of Economic Co-operation and Development.): In the absence of emissions trading between Annex B countries, the majority of global studies show reductions in projected GDP of about 0.2% to 2% in 2010 for different Annex II regions. With full emissions trading between Annex B countries, the estimated reductions in 2010 are between 0.1% and 1.1% of projected GDP. Models do not include sinks or non-CO_2 greenhouse gases. They do not include the CDM, negative cost options, ancillary benefits, or targeted revenue recycling. For all regions costs are also influenced by the following factors:

— Constraints on the use of Annex B trading, high transaction costs in implementing the mechanisms, and inefficient domestic implementation could raise costs.

— Inclusion in domestic policy and measures of the no regrets possibilities identified in Paragraph 12, use of the CDM, sinks, and inclusion of non-CO_2 greenhouse gases, could lower costs. Costs for individual countries can vary more widely.

The models show that the Kyoto mechanisms are important in controlling risks of high costs in given countries, and thus can complement domestic policy mechanisms. Similarly, they can minimise risks of inequitable international impacts and help to level marginal costs. The global modelling studies show national marginal costs to meet the Kyoto targets from about

US$20/tC up to US$600/tC without trading, and a range from about US$15/tC up to US$150/tC with Annex B trading. The cost reductions from these mechanisms may depend on the details of implementation, including the compatibility of domestic and international mechanisms, constraints, and transaction costs.

Economies in transition: For most of these countries, GDP effects range from negligible to a several per cent increase. This reflects opportunities for energy efficiency improvements not available to Annex II countries. Under assumptions of drastic energy efficiency improvement and/or continuing economic recessions in some countries, the assigned amounts may exceed projected emissions in the first commitment period. In this case, models show increased GDP due to revenues from trading assigned amounts. However, for some economies in transition, implementing the Kyoto Protocol will have similar impact on GDP as for Annex II countries.

Cost-effectiveness studies with a century timescale estimate that the costs of stabilising CO_2 concentrations in the atmosphere increase as the concentration stabilisation level declines. Different baselines can have a strong influence on absolute costs. While there is a moderate increase in the costs when passing from a 750ppmv to a 550ppmv concentration stabilisation level, there is a larger increase in costs passing from 550ppmv to 450ppmv unless the emissions in the baseline scenario are very low. These results, however, do not incorporate carbon sequestration, gases other than CO_2 and did not examine the possible effect of more ambitious targets on induced technological change.

Costs associated with each concentration level depend on numerous factors including the rate of discount, distribution of emission reductions over time, policies and measures employed, and particularly the choice of the baseline scenario: for scenarios characterised by a focus on local and regional sustainable development for example, total costs of stabilising at a particular level are significantly lower than for other scenarios. Under any greenhouse gas mitigation effort, the economic costs and benefits are distributed unevenly between sectors; to a varying degree, the costs of mitigation actions could be reduced by appropriate policies.

In general, it is easier to identify activities, which stand to suffer economic costs compared to those which may benefit, and the economic costs are more immediate, more concentrated and more certain. Under mitigation policies, coal, possibly oil and gas, and certain energy intensive sectors, such as steel production, are most likely to suffer an economic disadvantage. Other industries including renewable energy industries and services can be expected to benefit in the long term from price changes and the availability of financial and other resources that would otherwise have been devoted to carbon-intensive sectors.

Policies such as the removal of subsidies from fossil fuels may increase total societal benefits through gains in economic efficiency, while use of the Kyoto mechanisms could be expected to reduce the net economic cost of meeting Annex B targets. Other types of policies, for example exempting carbon-intensive industries, redistribute the costs but increase total societal costs at the same time. Most studies show that the distributional effects of a carbon tax can have negative income effects on low-income groups unless the tax revenues are used directly or indirectly to compensate such effects.

Emission constraints in Annex I countries have well established, albeit varied "spillover" effects on non-Annex I countries.

— *Oil-exporting, non-Annex I countries:* Analyses report costs differently, including, inter alia, reductions in projected GDP and reductions in projected oil revenues. The study reporting the lowest costs shows reductions of 0.2% of projected GDP with no emissions trading, and less than 0.05% of projected GDP with Annex B emissions trading in 2010. The study reporting the highest costs shows reductions of 25% of projected oil revenues with no emissions trading, and 13% of projected oil revenues with Annex B emissions trading in 2010. These studies do not consider policies and measures other than Annex B emissions trading, that could lessen the impact on non-Annex I, oil-exporting countries, and therefore tend to overstate both the costs to these countries and overall costs. The effects on these countries can be further reduced by removal of subsidies for fossil fuels, energy tax restructuring according to carbon content, increased use of natural gas, and diversification of the economies of non-Annex I, oil-exporting countries.

— *Other non-Annex I countries:* They may be adversely affected by reductions in demand for their exports to OECD nations and by the price increase of those carbon- intensive and other products they continue to import. These countries may benefit from the reduction in fuel prices, increased exports of carbon-intensive products and the transfer of environmentally sound technologies and know-how. The net balance for a given country depends on which of these factors dominates. Because of these complexities, the breakdown of winners and losers remains uncertain.

— *Carbon leakage:* The possible relocation of some carbon-intensive industries to non-Annex I countries and wider impacts on trade flows in response to changing prices may lead to leakage in the order of 5%-20%. Exemptions, for example for energy intensive industries, make the higher model estimates for carbon leakage unlikely, but would raise aggregate costs. The transfer of environmentally sound technologies and know-how, not included in models, may lead to lower leakage and especially on the longer term may more than offset the leakage.

Different Ways for Mitigation

The successful implementation of greenhouse gas mitigation options needs to overcome many technical, economic, political, cultural, social, behavioural and/or institutional barriers which prevent the full exploitation of the technological, economic and social opportunities of these mitigation options. The potential mitigation opportunities and types of barriers vary by region and sector, and over time. This is caused by the wide variation in mitigation capacity. The poor in any country are faced with limited opportunities to adopt technologies or change their social behaviour, particularly if they are not part of a cash economy, and most countries could benefit from innovative financing and institutional reform and removing barriers to trade.

In the industrialised countries, future opportunities lie primarily in removing social and behavioural barriers; in countries with economies in transition, in price rationalisation; and in developing countries, in price rationalisation, increased access to data and information, availability of advanced technologies, financial resources, and training and capacity building. Opportunities for any given country, however, might be found in the removal of any combination of barriers. National responses to climate change can be more effective if deployed as a portfolio of policy instruments to limit or reduce greenhouse gas emissions.

The portfolio of national climate policy instruments may include—according to national circumstances—emissions/carbon/energy taxes, tradable or non-tradable permits, provision and/or removal of subsidies, deposit/refund systems, technology or performance standards, energy mix requirements, product bans, voluntary agreements, government spending and investment, and support for research and development. Each government may apply different evaluation criteria, which may lead to different portfolios of instruments.

Market based instruments may be cost-effective in many cases, especially where capacity to administer them is developed. Energy efficiency standards and performance regulations are widely used, and may be effective in many countries, and sometimes precede market based instruments. Voluntary agreements have recently been used more frequently, sometimes preceding the introduction of more stringent measures. Information campaigns, environmental labelling, and green marketing, alone or in combination with incentive subsidies, are increasingly emphasised to inform and shape consumer or producer behaviour.

Government and/or privately supported research and development is important in advancing the long-term application and transfer of mitigation technologies beyond the current market or economic potential. The effectiveness of climate change mitigation can be enhanced when climate policies are integrated with the nonclimate objectives of national and sectorial policy development and be turned into broad transition strategies to achieve the long-term social and technological changes required by both sustainable development and climate change mitigation. Just as climate policies can yield ancillary benefits that improve wellbeing, non-climate policies may produce climate benefits. It may be possible to significantly reduce greenhouse gas emissions by pursuing climate objectives through general socioeconomic policies.

In many countries, the carbon intensity of energy systems may vary depending on broader programmmes for energy infrastructure development, pricing, and tax policies. Adopting state-of-the-art environmentally sound technologies may offer particular opportunity for environmentally sound development while avoiding greenhouse gas intensive activities. Specific attention can foster the transfer of those technologies to small and medium size enterprises. Moreover, taking ancillary benefits into account in comprehensive national development strategies can lower political and institutional barriers for climate-specific actions. Co-ordinated actions among countries and sectors may help to reduce mitigation cost, address competitiveness concerns, potential conflicts with international trade rules, and carbon leakage.

A group of countries that wants to limit its collective greenhouse gas emissions could agree to implement well-designed international instruments. Climate change decision-making is essentially a sequential process under general uncertainty. A prudent risk management strategy requires a careful consideration of the consequences, their likelihood and society's attitude toward risk.

The latter is likely to vary from country to country and perhaps even from generation to generation. Economic modelling studies completed since the SAR indicate that a gradual near-term transition from the world's present energy system towards a less carbon-emitting economy minimizes costs associated with premature retirement of existing capital stock. It also provides time for technology development, and avoids premature lockin to early versions of rapidly developing low-emission technology.

On the other hand, more rapid near-term action would decrease environmental and human risks associated with rapid climatic changes. It would also stimulate more rapid deployment of existing low emission technologies, provide strong near-term incentives to future technological changes that may help to avoid lock-in to carbon-intensive technologies, and allow for later tightening of targets should that be deemed desirable in light of evolving scientific understanding. There is an interrelationship between the environmental effectiveness of an international regime, the cost-effectiveness of climate policies and the equity of the agreement.

Any international regime can be designed in a way that enhances both its efficiency and its equity. Advances have been made since previous IPCC assessments in the understanding of the scientific, technical, environmental, and economic and social aspects of mitigation of climate change. Further research is required, however, to strengthen future assessments and to reduce uncertainties as far as possible in order that sufficient information is available for policy making about responses to climate change, including research in developing countries. The following are high priorities for further narrowing gaps between current knowledge and policy making needs:

— Further exploration of the regional, country and sector specific potentials of technological and social innovation options. This includes research on the short, medium and long-term potential and costs of both CO_2 and non-CO_2, non-energy mitigation options; understanding of technology diffusion across different regions; identifying opportunities in the area of social innovation leading to decreased greenhouse gas emissions; comprehensive analysis of the impact of mitigation measures on carbon flows in and out of the terrestrial system; and some basic inquiry in the area of geo-engineering.

— Economic, social and institutional issues related to climate change mitigation in all countries. Priority areas include: analysis of regionally specific mitigation options and barriers; the implications of equity assessments; appropriate methodologies and improved data sources for climate change mitigation and capacity building in the area of integrated assessment; strengthening future research and assessments, especially in the developing countries.

— Methodologies for analysis of the potential of mitigation options and their cost, with special attention to comparability of results. Examples include: characterising and measuring barriers that inhibit greenhouse gas-reducing action; making mitigation modelling techniques more consistent, reproducible, and accessible; modelling technology learning; improving analytical tools for evaluating ancillary benefits, e.g. assigning the costs of abatement to greenhouse gases and to other pollutants; systematically analysing the dependency of costs on baseline assumptions for various greenhouse gas stabilisation scenarios; developing decision analytical frameworks for dealing with uncertainty as well as socio-economic and ecological risk in climate policy making; improving global models and studies, their assumptions and their consistency in the treatment and reporting of non-Annex I countries and regions.

— Evaluating climate mitigation options in the context of development, sustainability and equity. Examples include: exploration of alternative development paths, including sustainable consumption patterns in all sectors, including the transportation sector; integrated analysis of mitigation and adaptation; identifying opportunities for synergy between explicit climate policies and general policies promoting sustainable development; integration of intra- and inter-generational equity in climate change mitigation analysis; implications of equity assessments; analysis of scientific, technical and economic implications of options under a wide variety of stabilisation regimes.

11

Protecting the Ozone Layer

Now a days ozone-depleting substances (ODSs) affect the global climate system very well. ODSs cause depletion of the stratospheric ozone layer, their production and consumption are controlled with efforts made by both developed and developing countries. Both the ODSs and a number of their substitutes are greenhouse gases (GHGs) which contribute to climate change.

Climate change may also indirectly influence the ozone layer. The major application sectors using ODSs and their hydrofluorocarbons (HFCs) and perfluorocarbons (PFCs) substitutes include refrigeration, air conditioning, foams, aerosols, fire protection and solvents. Emissions of these substances originate from manufacture and any unintended by-product releases, intentionally emissive applications, evaporation and leakage from banks contained in equipment and products during use, testing and maintenance, and end of-life practices.

Hydrochlorofluorocarbons (HCFCs) have been used to replace chlorofluorocarbons (CFCs) in several applications as they have shorter lifetimes in the atmosphere and consequently cause less ozone depletion. HFCs and PFCs have been identified as potential long-term replacements for ODSs because they contain neither

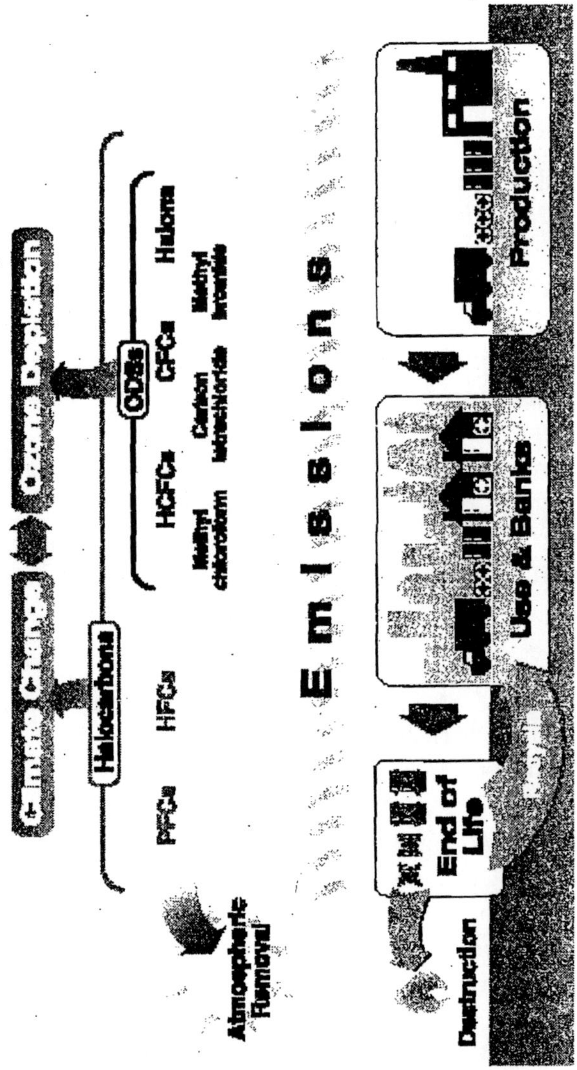

Figure 1. Schematic diagram of major factors in affect the climatic changes

bromine nor chlorine and do not cause any significant ozone depletion. However, all of these species are also GHGs and so contribute to climate change in varying degrees. Additional alternatives for halocarbon use include ammonia and organic substances, the direct emissions of which have a very small effect on climate although indirect emissions may be important.

The major application sectors using ODSs and their HFC/ PFC substitutes include refrigeration, air conditioning, foams, aerosols, fire protection and solvents. Emissions of these substances originate from manufacture and any unintended byproduct releases, intentionally emissive applications, evaporation and leakage from banks contained in equipment and products during use, testing and maintenance, and end of-life practices. With regard to specific emission reduction options, the report generally limits its coverage to the period up to 2015, for which reliable literature is available on replacement options with significant market potential for these rapidly evolving sectors.

Many halocarbons, including CFCs, PFCs, HFCs and HCFCs, are effective GHGs because they absorb Earth's outgoing infrared radiation in a spectral range where energy is not removed by CO_2 or water vapour. Halocarbon molecules can be many thousand times more efficient at absorbing radiant energy emitted from the Earth than a molecule of CO_2, and small amounts of these gases can contribute significantly to radiative forcing of the climate system. For most of the species considered here, the magnitude of the direct radiative forcing generated by a gas is given by the product of its mixing ratio (in parts per billion, ppb) and radiative efficiency.

For the more abundant greenhouse gases CO_2, methane and nitrous oxide there is a nonlinear relationship between mixing ratio and radiative forcing. The primary radiative effect of CO_2 and water vapour is to warm the surface climate but cool the stratosphere. However, due to their absorption in the atmospheric window, the direct radiative effect of halocarbons is to warm both the troposphere and stratosphere.

Apart from their direct effect on climate, some gases have an indirect effect either from radiative forcing caused by their degradation products or through their influences on atmospheric chemistry. Halocarbons containing chlorine and bromine are ODSs and, because ozone is a strong absorber of UV radiation, they have an indirect cooling effect which is significant compared to their direct warming effect. For hydrocarbons, indirect effects associated with tropospheric ozone production may be significantly greater than direct effects.

Effects of ODSs and their Substitutes

Halocarbons, and in particular ODSs, have contributed to positive direct radiative forcing and associated increases in global average surface temperature. The total positive direct radiative forcing due to increases in industrially produced ODS and non-ODS halocarbons from 1750 to 2000 is estimated to be 0.33 ± 0.03 W m^{-2}, representing about 13% of the total due to increases in all well-mixed greenhouse gases over that period. Most halocarbon increases have occurred in recent decades. Atmospheric concentrations of CFCs (chlorofluorocarbons) were stable or decreasing in the period 2001-2003 (0 to -3% per year, depending on the specific gas) while the halons and the substitute hydrochlorofluorocarbons (HCFCs) and HFCs increased (+1 to +3% per year, +3 to +7% per year, and +13 to +17% per year, respectively).

Stratospheric ozone depletion observed since 1970 is caused primarily by increases in concentrations of reactive chlorine and bromine compounds that are produced by degradation of anthropogenic ODSs, including halons, CFCs, HCFCs, methyl chloroform (CH_3CCl_3), carbon tetrachloride (CCl_4) and methyl bromide (CH_3Br). Ozone depletion produces a negative radiative forcing of climate, which is an indirect cooling effect of the ODSs. Changes in ozone are believed to currently contribute a globally averaged radiative forcing of about -0.15 ± 0.10 W m^{-2}.

The large uncertainty in the indirect radiative forcing of ODSs arises mainly because of uncertainties in the detailed vertical distribution of ozone depletion. This negative radiative forcing is very likely to be smaller than the positive direct radiative forcing due to ODSs alone (0.32 ± 0.03 W m^{-2}). Warming due to ODSs and cooling associated with ozone depletion are two distinct climate forcing mechanisms that do not simply offset one another. The spatial and seasonal distributions of the cooling effect of ozone depletion differ from those of the warming effect.

A limited number of global climate modelling and statistical studies suggest that ozone depletion is one mechanism that may affect patterns of climate variability which are important for tropospheric circulation and temperatures in both hemispheres. However, observed changes in these patterns of variability cannot be unambiguously attributed to ozone depletion. Each type of gas has had different greenhouse warming and ozone depletion effects depending mainly on its historic emissions, effectiveness as a greenhouse gas, lifetime and the amount of chlorine and/ or bromine in each molecule. Bromine-containing gases currently contribute much more to cooling than to warming, whereas CFCs and HCFCs contribute more to warming than to cooling. HFCs and PFCs contribute only to warming.

Actions taken under the Montreal Protocol have led to the replacement of CFCs with HCFCs, HFCs, and other substances and processes. Because replacement species generally have lower global warming potentials (GWPs), and because total halocarbon emissions have decreased, their combined CO_2-equivalent (direct GWP-weighted) emission has been reduced. The combined CO_2-equivalent emissions of CFCs, HCFCs and HFCs derived from atmospheric observations decreased from about 7.5 ± 0.4 $GtCO_2$-eq per year around 1990 to 2.5 ± 0.2 $GtCO_2$-eq per year around 2000, equivalent to about 33% and 10%, respectively, of the annual CO_2 emissions due to global fossil fuel burning.

Stratospheric chlorine levels have approximately stabilized and may have already started to decline. Ammonia and those hydrocarbons (HCs) used as halocarbon substitutes have atmospheric lifetimes ranging from days to months, and the direct and indirect radiative forcings associated with their use as substitutes are very likely to have a negligible effect on global climate. Changes in energy-related emissions associated with their use may also need to be considered.

According to the IPCC Special Report on Emission Scenarios (SRES), the direct radiative forcing of HFCs in 2015 is about 0.030 W m^{-2}, and the radiative forcing of PFCs in 2015 is about 0.006 W m^{-2}. Those HFC and PFC radiative forcings correspond to about 1.0% and 0.2%, respectively, of the estimated radiative forcing of all well-mixed greenhouse gases in 2015, with the contribution of ODSs being about 10%. While this report particularly focused on scenarios for the period up to 2015, for the period beyond 2015 the IPCC SRES scenarios were considered but were not re-assessed. These SRES scenarios project significant growth in radiative forcing from HFCs over the following decades, but the estimates are likely to be very uncertain due to growing uncertainties in technological practices and policies.

Observations and model calculations suggest that the global average amount of ozone depletion has now approximately stabilised. Although considerable variability in ozone is expected from year to year, including in polar regions where depletion is largest, the ozone layer is expected to begin to recover in coming decades due to declining ODS concentrations, assuming full compliance with the Montreal Protocol. Over the long term, projected increases in other greenhouse gases could increasingly influence the ozone layer by cooling the stratosphere and changing stratospheric circulation.

As a result of the cooling effect and of reducing ODS concentrations, ozone is likely to increase over much of the stratosphere, but could decrease in some regions, including the Arctic. However, the effects of changes in atmospheric circulation associated with climate change could be larger than these factors, and the net impact on total ozone due to increases in atmospheric concentrations of greenhouse gases is currently uncertain in both magnitude and sign. Based on current models an Arctic 'ozone hole' similar to that presently observed over the Antarctic is very unlikely to occur.

The relative future warming and cooling effects of emissions of CFCs, HCFCs, HFCs, PFCs and halons vary with gas lifetimes, chemical properties and time of emission. The atmospheric lifetimes range from about a year to two decades for most HFCs and HCFCs, decades to centuries for some HFCs and most halons and CFCs, and 1000 to 50,000 years for PFCs. Direct GWPs for halocarbons range from 5 to over 10,000. ODS indirect cooling is projected to cease upon ozone layer recovery, so that GWPs associated with the indirect cooling effect depend on the year of emission, compliance with the Montreal Protocol and gas lifetimes. These indirect GWPs are subject to much greater uncertainties than direct GWPs.

Impacts on Global Tropospheric Chemistry

Substitution for ODSs in air conditioning, refrigeration, and foam blowing by HFCs, PFCs, and other gases such as hydrocarbons are not expected to have a significant effect on global tropospheric chemistry. Small but not negligible impacts on air quality could occur near localized emission sources and such impacts may be of some concern, for instance in areas that currently fail to meet local standards.

Persistent degradation products of HFCs and HCFCs are removed from the atmosphere via deposition and washout processes. However, existing environmental risk assessment and monitoring studies indicate that these are not expected to result in environmental concentrations capable of causing significant ecosystem damage. Measurements of TFA (trifluoroacetic acid) in sea water indicate that the anthropogenic sources of TFA are smaller than natural sources, but the natural sources are not fully identified.

Lifetimes of CFCs in the Atmosphere

Gases with longer lifetimes remain in the atmosphere, and consequently may affect climate, for a longer time. Most CFCs are removed from the atmosphere on time scales ranging from about 50 to 100 years. With the exception of HFC-23, which has a lifetime of 270 years, HCFCs and HFCs are removed efficiently in the troposphere through atmospheric chemistry oxidation processes. As a result, they have lifetimes ranging from about one year to a few decades. PFCs are very inert molecules and their emissions contribute to warming of the climate system over timescales that can exceed 1000 years.

Most halocarbon gases are sufficiently long-lived that they become mixed throughout the atmosphere before they are destroyed. They therefore have nearly constant mixing ratios throughout the troposphere. In contrast, ammonia and organic compounds have lifetimes that typically range from days to weeks, thereby making their distributions variable both spatially and temporally.

Atmospherisc Concentration of Halocarbons

Currently observed rates of change differ among ODSs, depending mainly upon their emissions and atmospheric lifetimes. Following the Montreal Protocol and its subsequent amendments, large reductions in ODS production and emission have occurred. The concentrations of some CFCs have peaked, while others are expected to decline in the future. In the complete absence of emissions, the concentrations of these gases would decrease, but at a slow rate that is determined by their atmospheric lifetimes of decades to centuries.

Continuing emissions will cause increases in the concentrations of these gases or diminish their rates of decrease. Observations of annual changes in atmospheric concentrations provide the most reliable estimates of total global emissions for long-lived gases. For example, CFC-11 is observed to be decreasing at a rate about 60% slower than would occur in the absence of emissions, while CFC-12 is still increasing slightly, indicating continuing emissions of both species. CFC-113, however, is decreasing at a rate close to that expected in the absence of emissions.

For comparison, earlier estimates of emissions are also shown for the year 1990, which is slightly after the peak in ODS emissions. Emissions of CFC-113 and CH_3CCl_3, both used largely as solvents and having no accumulated banks, decreased by more than a factor of 10 from 1990 to 2000. Stratospheric chlorine levels have approximately stabilised and may have already started to decline. In contrast, production is currently greater than emission for nearly all of the HCFCs and HFCs, implying that banks of these chemicals are currently accumulating and could contribute to future radiative forcing. One measure of the relevance of such banks is the ratio of bank size to the amount already in the atmosphere. For example, in the case of HFC-134a these amounts are estimated to be about equal.

Continuing atmospheric observations of CFCs and other ODSs now enable improved validation of estimates for the lag between production and emission to the atmosphere. This provides new insight into the overall significance of banks and of end-of-life options which are relevant to the future use of HCFC and HFC substitutes. For some gases, there are now sufficient atmospheric observations to constrain not just global but also regional emissions in certain areas. For example, atmospheric measurements suggest sharp increases in European emissions of HFC-134a over the period 1995-1998 and in HFC-152a over the period 1996-2000, with some subsequent levelling off through to 2003.

The direct radiative forcing due to these increases from 1750 to 2000 is estimated to be 0.33 ± 0.03 W m^{-2}, which represents about 13% of the total due to increases in all of the well-mixed GHGs over that period. The contributions of CFCs, HCFCs and HFCs are about 0.27 W m^{-2}, 0.033 W m-2 and 0.006 W m^{-2} respectively. Because increases in halocarbon concentrations occurred mainly during the last three decades, their relative contribution to total radiative forcing is larger during this period. The direct radiative forcing due to increases in halocarbons from 1970 to 2000 was 0.27 ± 0.03 W m^{-2}, which represents about 23% of that due to increases in all of the well-mixed GHGs. The contribution to direct radiative forcing due to HCFCs is presently dominated by HCFC-22, while that due to HFCs is dominated by HFC-134a and HFC-23, with the latter being emitted mainly as a byproduct of manufacture of HCFC-22.

Cause of Stratospheric Ozone Change

The largest decreases since 1980 have been observed over the Antarctic during the spring (the Antarctic ozone hole), with the monthly total column ozone amounts in September and October being about 40-50% below preozone-hole values.

Arctic ozone loss in any given year depends strongly on the meteorological conditions. Arctic ozone has been chemically depleted by up to 30% in recent cold years, but the losses observed in warm years have been very small. Globally averaged ozone has decreased by roughly 3% since 1980. The ozone column decreased by about 6% over the midlatitudes (35°-60°) in the Southern Hemisphere and 3% in the Northern Hemisphere. Significant long-term changes in column ozone have not been observed in the tropics. Observations and model calculations suggest that the global average amount of ozone depletion has now approximately stabilised.

The observed ozone changes are caused by both chemical and dynamical factors, with the chemical factors being dominant. This ozone depletion is caused primarily by increases in concentrations of reactive chlorine and bromine compounds that are produced by degradation of anthropogenic ODSs, including halons, CFCs, HCFCs, methyl chloroform (CH_3CCl_3), carbon tetrachloride (CCl_4) and methyl bromide (CH_3Br). Human activities have increased the amount of chlorine in the stratosphere, relative to estimated natural background levels, by a factor of about 5 since 1970. CFCs are the primary source of this change, while HCFCs currently contribute about 5% to the total stratospheric chlorine loading.

Effects of Ozone Depletion

Observations and modelling show that ozone depletion has acted to cool the stratosphere, which in turn can contribute to cooling of the troposphere and surface. The warming of the climate by ODSs and the cooling associated with ozone depletion are two distinct mechanisms that are governed by different physical processes and feedbacks and for which there are quite different levels of scientific understanding.

Gases containing bromine are particularly effective ozone depleters and have a larger contribution to the indirect effect on a per-molecule basis than other ozone-depleting gases such as the CFCs. The best estimate of the negative indirect radiative forcing associated with ozone depletion over the period 1970-2000 is -0.15 ± 0.10 W m^{-2}, where the large uncertainty is determined by the range of model estimates and arises mainly because of uncertainties in the detailed vertical distribution of ozone depletion. This indirect effect is very likely to be smaller in magnitude than the positive direct radiative forcing due to ODSs alone (0.32 ± 0.03 W m^{-2}) which is far better understood.

If some fraction of the observed global ozone changes were not attributable to ODSs the magnitude of this indirect effect would be reduced. A limited number of global climate modelling and statistical studies suggest that ozone depletion is one mechanism that may affect patterns of climate variability which are important for tropospheric circulation and temperatures in both hemispheres. In particular, the significant depletion of stratospheric ozone occurring in the Antarctic region is likely to have affected stratospheric circulation, and consequently the troposphere. There are indications that the Antarctic ozone hole has contributed to the cooling observed over the

Antarctic plateau and to warming in the region of the Antarctic Peninsula.

Factors for Controlling Ozone

Global ozone recovery is expected to follow decreases in chlorine and bromine loading in the stratosphere as ODS concentrations decline due to reductions in their emissions. While this is expected to be the dominant factor in ozone recovery, emissions of other GHGs can affect both tropospheric and stratospheric chemistry and climate, and will have some effect on ozone recovery. Future temperature changes related to GHG emissions are expected to enhance stratospheric ozone depletion in some parts of the stratosphere and decrease it in others.

Increases in CO_2 concentration are expected to cool the stratosphere, which is known to reduce the rates of gasphase ozone destruction in much of the stratosphere and thereby increase ozone concentrations at altitudes above about 25 km. In contrast, lower temperatures could decrease ozone concentrations at lower altitudes. While the latter effect is expected to be most important in the Arctic in late winter to early spring, it may be small compared with other processes and will slowly decrease with time as the chlorine and bromine loadings decrease.

Changes in stratospheric circulation may also occur in association with increases in GHGs and these could either increase or decrease future mid-latitude and polar ozone. The net result of future GHG emissions on global ozone depends upon the combination of these effects and its magnitude and direction are poorly quantified at present. Models suggest that minimum levels of Antarctic ozone may already have occurred or should occur within the next few years. Predictions of the timing of a minimum in Arctic ozone are more uncertain due to far greater natural variability in this region, but models suggest that it should occur within the next two decades. An Arctic 'ozone hole' similar to that currently observed over the Antarctic is very unlikely to occur.

The estimated radiative forcing of HFCs in 2015 is in the range 0.022-0.025 W m^{-2} based on the SRES emission scenarios and in the range 0.019-0.030 W m^{-2}. The radiative forcing of PFCs in 2015 is about 0.006 W m^{-2} based on SRES scenarios. These HFC and PFC radiative forcings correspond to about 6-10% and 2% respectively of the total estimated radiative forcing due to CFCs and HCFCs in 2015. The magnitude of current ODS banks remains uncertain, their contribution to radiative forcing is expected to be comparable with that of HFC emissions to the atmosphere in the next few decades. This indicates that choices with respect to end-of-life options, such as the destruction of currently banked material, can provide significant benefits for the climate system.

In addition to reducing atmospheric chlorine loading, actions taken under the Montreal Protocol and its adjustments and amendments have also acted to reduce total CO_2-equivalent emissions. This has occurred because the replacement species generally have lower global warming potentials (GWPs) and because total halocarbon emissions have been reduced. Ammonia and those hydrocarbons used as halocarbon substitutes have atmospheric lifetimes ranging from days to months, and the direct and indirect radiative forcings associated with their use as substitutes are very likely to have a negligible effect on global climate.

Global Warming Potentials (GWPs)

The GWP is a measure of the future radiative effect of an emission of a substance relative to the emission of the same amount of CO_2 integrated over a chosen time horizon. The GWP value for a species is proportional to its radiative efficiency and increases with its lifetime in the atmosphere. GWPs are most useful as relative measures of the climate response due to direct radiative forcing of well-mixed greenhouse gases whose atmospheric lifetimes are controlled by similar processes, which includes most of the halocarbons. The choice of time horizon is a policy consideration and makes a compromise between the relative weighting of short-term and long-term effects.

Current practice most often uses GWPs calculated with a 100-year time horizon. This does not take into account the full effect of very long-lived gases with high radiative efficiency, such as PFCs, that persist in the atmosphere for much longer than 100 years. Similarly, integrating over 100 years reduces the contribution of short-lived species that last for only part of that period, relative to that of CO_2 which continues to contribute to radiative forcing throughout the 100-year time horizon and beyond. As ODS indirect cooling effects are projected to cease upon ozone layer recovery, their duration depends not only on gas lifetime but also on the time of ozone recovery.

Estimates of indirect GWPs incorporate this latter dependence by setting all indirect effects to zero after the time at which equivalent effective stratospheric chlorine (EESC) is estimated to return to its pre-1980 values. Indirect GWPs therefore depend on the year of emission and have large uncertainties arising from: uncertainty in the radiative forcing caused by ozone depletion; uncertainties in the amount of EESC attributed to each species; and uncertainty in the time at which EESC returns to its pre-1980 values. Where direct GWPs are used with ODS emissions, or to construct CO_2-equivalent values, it should be recognised that there are also indirect effects that may be significant over the next several decades.

Production, Banks and Emissions

Current emissions of ODSs and their substitutes are largely determined by historic use patterns. For CFCs and HCFCs, a significant contribution comes from their respective banks. There are no regulatory obligations to restrict these CFC and HCFC emissions either under the Montreal Protocol and its Kyoto Protocol, although some countries have effective national policies for this purpose. Banks are the total amount of substances contained in existing equipment, chemical stockpiles, foams and other products not yet released to the atmosphere. The buildup of banks of (relatively) new applications of HFCs will—in the absence of additional bank management measures—also significantly determine post 2015 emissions.

Observations of atmospheric concentrations, combined with production and use pattern data, can indicate the significance of banks, but not their exact sizes. The most accurate estimates of emissions of CFC-11 and CFC-12 are derived from observations of atmospheric concentrations. Those emissions are now larger than estimated releases related to current production, indicating that a substantial fraction of these emissions come from banks built up through past production.

Observations of atmospheric concentrations show that global emissions of HFC-134a are presently smaller than reported production, implying that this bank is growing. The total global amount of HFC-134a currently in the atmosphere is believed to be about equal to the amount in banks. In the case of CFC-11 and some other gases, the lack of information on use patterns makes it difficult to assess the contribution to observed emissions from current production and use. Further work in this area is required to clarify the sources.

Banks of CFCs, HCFCs, HFCs and PFCs were estimated at about 21 $GtCO_2$-eq in 2002. In a Business-As-Usual (BAU) scenario, banks are projected to decline to about 18 $GtCO_2$-eq in 2015. In 2002, CFC, HCFC and HFC banks were about 16, 4 and 1 $GtCO_2$-eq, respectively. In 2015, the banks are about 8, 5 and 5 $GtCO_2$-eq, respectively, in the BAU scenario. Banks of PFCs used as ODS replacements were about 0.005 $GtCO_2$-eq in 2002. CFC banks associated with refrigeration, stationary airconditioning (SAC) and mobile air-conditioning (MAC) equipment are projected to decrease from about 6 to 1 $GtCO_2$-eq over the period 2002 to 2015, mainly due to release to the atmosphere and partly due to end-of-life recovery and destruction.

CFC banks in foams are projected to decrease much more slowly over the same period (from 10 to 7 $GtCO_2$-eq), reflecting the much slower release of banked blowing agents from foams when compared with similarly sized banks of refrigerant in the refrigeration and airconditioning sector. HFC banks have started to build up and are projected to reach about 5 $GtCO_2$-eq in 2015. Of these, HFCs banked in foams represent only 0.6 $GtCO_2$-eq, but are projected to increase further after 2015. In the BAU scenario, total direct emissions of CFCs, HCFCs, HFCs and PFCs are projected to represent about 2.3 $GtCO_2$-eq per year by 2015.

CFC and HCFC emissions are together decreasing from 2.1 to 1.2 $GtCO_2$-eq per year, and emissions of HFCs are increasing from 0.4 to 1.2 $GtCO_2$-eq per year. PFC emissions from ODS substitute use are about 0.001 $GtCO_2$-eq per year and projected to decrease. Refrigeration applications together with SAC and MAC contribute the bulk of global direct GHG emissions in line with the higher emission rates associated with refrigerant banks. The largest part of GHG emissions from foams is expected to occur after 2015 because most releases occur at end-of-life. With little new production, total CFC banks will decrease due to release to the atmosphere during operation and disposal.

In the absence of additional measures a significant part of the CFC banks will have been emitted by 2015. Consequently, annual CFC emissions are projected to decrease from 1.7 to 0.3 $GtCO_2$-eq per year. HCFC emissions are projected to increase from 0.4 to 0.8 $GtCO_2$-eq per year, owing to a steep increase expected for their use in (commercial) refrigeration and SAC applications. The projected threefold increase in HFC emissions is the result of increased application of HFCs in the refrigeration, SAC and MAC sectors, and due to byproduct emissions of HFC-23 from increased HCFC-22 production.

Comparison of results of atmospheric measurements with inventory calculations shows differences per group of substances in the order of 10 to 25%. For individual gases the differences can be much bigger. This is caused by unidentified emissive applications of some substances, not accounted for in inventory calculations, and uncertainties in the geographically distributed datasets of equipment in use. For individual applications, the relevance of indirect GHG emissions over a life cycle can range from low to high, and for certain applications may be up to an order of magnitude larger than direct GHG emissions. This is highly dependent on the specific sector and product/ application characteristics, the carbon-intensity of the consumed electricity and fuels during the complete life cycle of the application, containment during the use-phase, and the end-of-life treatment of the banked substances.

Methods for Reducing Greenhouse Gas Emissions

Methods for determining which technology option has the highest GHG emission reduction potential address both direct emissions of halocarbons or substitutes and indirect energy-related emissions over the full life cycle. In addition, comprehensive methods assess a wide range of environmental impacts. Other, simplified methods exist to assess life-cycle impacts and commonly provide useful indicators for life-cycle greenhouse gas emissions of an application. The conclusions from these comparisons are sensitive to assumptions about application-specific, and often region- and time-specific parameters (e.g., site-specific situation, prevailing climate, energy system characteristics).

Comparative economic analyses are important to identify cost-effective reduction options. However, they require a common set of methods and assumptions (e.g., costing methodology, time-frame, discount rate, future economic conditions, system boundaries). The development of simplified standardised methodologies would enable better comparisons in the future. The risks of health and safety impacts can be assessed in most cases using standardised methods.

Energy efficiency improvements can lead to reductions in indirect emissions from these appliances, depending on the particular energy source used and other circumstances, and produce net cost reductions, particularly where the use-phase of the application is long (e.g., in refrigeration and SAC). Through application of current best practices and recovery methods, there is potential to halve (1.2 $GtCO_2$-eq per year reduction) the BAU direct emissions from ODSs and their GHG substitutes by 2015. About 60% of this potential concerns HFC emissions, 30% HCFCs and 10% CFCs.

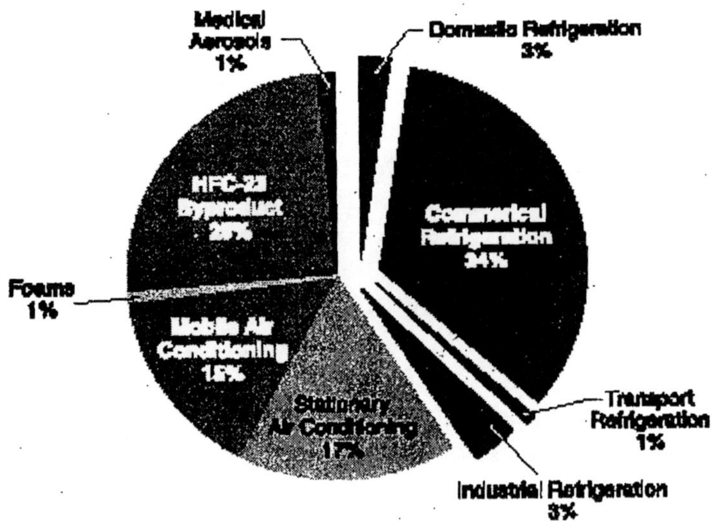

Figure 2. Sectirial Emission Reduction Potentials in 2015

The estimates are based on a Mitigation Scenario which makes regionally differentiated assumptions on best practices in production, use, substitution, recovery and destruction of these substances. Sectoral contributions are shown in Figure 2.

Sectoral Emission Reduction Potentials in 2015

In refrigeration applications direct GHG emissions can be reduced by 10% to 30%. For the refrigeration sector as a whole, the Mitigation Scenario shows an overall direct emission reduction of about 490 MtCO_2-eq per year by 2015, with about 400 MtCO_2-eq per year predicted for commercial refrigeration. Specific costs are in the range of 10 to 300 US$/t$CO_2$-eq. Improved system energy efficiencies can also significantly reduce indirect GHG emissions.

In full supermarket systems, up to 60% lower LCCP values can be obtained by using alternative refrigerants, improved containment, distributed systems, indirect systems or cascade systems. Refrigerant specific emission abatement costs range for the commercial refrigeration sector from 20 to 280 US$/ t$CO_2$-eq. In food processing, cold storage and industrial refrigeration, ammonia is forecast for increased use in the future, with HFCs replacing HCFC-22 and CFCs. Industrial refrigeration refrigerant specific emissions abatement costs were determined to be in the range from 27 to 37 US/tCO_2-eq. In transport refrigeration, lower GWP alternatives, such as ammonia, hydrocarbons and ammonia/carbon dioxide have been commercialised.

The emission reduction potential in domestic refrigeration is relatively small, with specific costs in the range of 0 to 130 US$/t$CO_2$-eq. Indirect emissions of systems using either HFC-134a or HC-600a (isobutane) dominate total emissions, for different carbon intensities of electric power generation. The difference between the LCCP of HFC-134a and isobutane systems is small and end-of-life recovery, at a certain cost increase, can further reduce the magnitude of the difference. Direct GHG emissions of residential and commercial airconditioning and heating equipment (SAC) can be reduced by about 200 MtCO_2-eq per year by 2015 relative to the BAU scenario.

Specific costs range from -3 to 170 US$/ t$CO_2$-eq. When combined with improvements in system energy efficiencies, which reduce indirect GHG emissions, in many cases net financial benefits accrue. Opportunities to reduce direct GHG (i.e., refrigerant) emissions can be found in (i) more efficient recovery of refrigerant at end of-life (in the Mitigation Scenario assumed to be 50% and 80% for developing and developed countries, respectively);

(ii) refrigerant charge reduction (up to 20%); (iii) better containment and (iv) the use of refrigerants with reduced or negligible GWPs in suitable applications.

Improving the integrity of the building envelope (reduced heat gain or loss) can have a significant impact on indirect emissions. HFC mixtures and hydrocarbons (HCs) (for small systems) are used as alternatives for HCFC-22 in developed countries. For those applications where HCs can be safely applied, the energy efficiency is comparable to fluorocarbon refrigerants. Future technical developments could reduce refrigerant charge, expanding the applicability of HCs. In mobile air conditioning, a reduction potential of 180 $MtCO_2$-eq per year by 2015 could be achieved at a cost of 20 to 250 US$/$tCO_2$-eq. Specific costs differ per region and per solution.

Improved containment, and end-of-life recovery and recycling could reduce direct GHG emissions by up to 50%, and total (direct and indirect) GHG emissions of the MAC unit by 30 to 40% at a financial benefit to vehicle owners. New systems with either CO_2 or HFC-152a, with equivalent LCCP, are likely to enter the market, leading to total GHG system emission reductions estimated at 50 to 70% in 2015 at an estimated added specific cost of 50 to 180 US$ per vehicle. Hydrocarbons and hydrocarbon blends, which have been used to a limited extent, present suitable thermodynamic properties and permit high energy efficiency.

However, the safety and liability concerns identified by vehicle manufacturers and suppliers limit the possible use of hydrocarbons in new vehicles. Due to the long life-span of most foam applications, by 2015 a limited emission reduction of 15 to 20 $MtCO_2$-eq per year is projected at specific costs ranging from 10 to 100 US$/$tCO_2$-eq. The potential for emission reduction increases in following decades. Several short-term emission reduction steps, such as the planned elimination of HFC use in emissive one-component foams in Europe, are already in progress and are considered as part of the BAU. Two further key areas of potential emission reduction exist in the foams sector.

The first is a potential reduction in halocarbon use in newly manufactured foams. However, the enhanced use of blends and the further phaseout of fluorocarbon use both depend on further technology development and market acceptance. Actions to reduce HFC use by 50% between 2010 and 2015, would result in emission reduction of about 10 $MtCO_2$-eq per year, at a specific cost of 15 to 100 US\$/tCO_2-eq, with further reductions thereafter.

The second opportunity for emission reduction can be found in the worldwide banks of halocarbons contained in insulating foams in existing buildings and appliances. Although recovery effectiveness is yet to be proven, and there is little experience to date, particularly in the buildings sector, commercial operations are already recovering halocarbons from appliances at 10 to 50 US\$/tCO_2-eq. Emission reductions may be about 7 $MtCO_2$-eq per year in 2015. However, this potential could increase significantly in the period between 2030 and 2050, when large quantities of building insulation foams will be decommissioned.

The reduction potential for medical aerosols is limited due to medical constraints, the relatively low emission level and the higher costs of alternatives. The major contribution to a reduction of GHG emissions for metered dose inhalers (MDIs) would be the completion of the transition from CFC to HFC MDIs beyond what was already assumed as BAU. The health and safety of the patient is considered to be of paramount importance in treatment decisions, and there are significant medical constraints to limit the use of HFC MDIs.

If salbutamol MDIs would be replaced by dry powder inhalers this would result in an annual emission reduction of about 10 $MtCO_2$-eq per year by 2015 at projected costs in the range of 150 to 300 US\$/tCO_2-eq. In fire protection, the reduction potential by 2015 is small due to the relatively low emission level, the significant shifts to not-in-kind alternatives in the past and the lengthy procedures for introducing new equipment. Direct GHG emissions for the sector are estimated at about 5 $MtCO_2$-eq per year in 2015 (BAU). Seventy five percent of original halon use has been shifted to agents with no climate impact.

Four percent of the original halon applications continue to employ halons. The remaining 21% has been shifted to HFCs with a small number of applications shifted to HCFCs and to PFCs. PFCs are no longer needed for new fixed systems and are limited to use as a propellant in one manufacturer's portable extinguisher agent blend. Due to the lengthy process of testing, approval and market acceptance of new fire protection equipment types and agents, no additional options will likely have appreciable impact by 2015.

With the introduction of a fluoroketone (FK) in 2002, additional reductions at an increased cost are possible in this sector through 2015. Currently those reductions are estimated to be small compared to other sectors. For non-medical aerosols and solvents there are several reduction opportunities, but the reduction potentials are likely to be rather small because most remaining uses are critical to performance or safety. The projected BAU emissions by 2015 for solvents and aerosols are about 14 and 23 $MtCO_2$-eq per year, respectively.

Substitution of HFC-134a by HFC-152a in technical aerosol dusters is a leading option for reducing GHG emissions. For contact cleaners and plastic casting mould release agents, the substitution of HCFCs by hydrofluoroethers (HFEs) and HFCs with lower GWPs offers an opportunity. Some countries have banned HFC use in cosmetic, convenience and novelty aerosol products, although HFC-134a continues to be used in many countries for safety reasons.

A variety of organic solvents can replace HFCs, PFCs and ODSs in many applications. These alternative fluids include lower GWP compounds such as traditional chlorinated solvents, HFEs, HCs and oxygenated solvents. Many not-inkind technologies, including no-clean and aqueous cleaning processes, are also viable alternatives. Destruction of byproduct emissions of HFC-23 from HCFC-22 production has a reduction potential of up to 300 $MtCO_2$-eq per year by 2015 and specific costs below 0.2 US$/$tCO_2$-eq according to two European studies in 2000. Reduction of HCFC-22 production due to market forces or national policies, or improvements in facility design and construction also could reduce HFC-23 emissions.

Policies, Measures and Instruments for Reducing ODSs Emission

A variety of policies, measures and instruments have been implemented in reducing the use or emissions of ODSs and their substitutes, such as HFCs and PFCs. These include regulations, economic instruments, voluntary agreements and international cooperation. Furthermore, general energy or climate policies affect the indirect GHG emissions of applications with ODSs, their substitutes, or not-in-kind alternatives. The relevant policies and approaches include:

— Regulations (e.g., performance standards, certification, restrictions, end-of-life management)

- Economic instruments (e.g., taxation, emissions trading, financial incentives and deposit refunds)
- Voluntary agreements (e.g., voluntary reductions in use and emissions, industry partnerships and implementation of good practice guidelines)
- International cooperation (e.g., Clean Development Mechanism)

It should be noted that policy considerations are dependent on specific applications, national circumstances and other factors.

HFC and PFC Production in the Future

There are no technical or legal limits to HFC and PFC production, it can be assumed that the global production capacity will generally continue to satisfy or exceed demand. In the BAU scenario, global production capacity is expected to expand with additions taking place mainly in developing countries and through joint ventures. Global production capacity of HFCs and PFCs most often exceeds current demand. There are a number of HFC-134a plants in developed countries and one plant in a developing country with others planned; the few plants for other HFCs are almost exclusively in developed countries.

The proposed European Community phase-out of HFC-134a in mobile air conditioners in new cars and the industry voluntary programme to reduce their HFC-134a emissions by 50% will affect demand and production capacity and output. Rapidly expanding markets in developing countries, in particular for replacements for CFCs, is resulting in new capacity for fluorinated gases which is at present being satisfied through the expansion of HCFC-22 and 141b capacity.

Emissions of GHG

Ozone-depleting substances (ODSs) and their substitutes are being used in a wide range of products and processes. Many of these substances (or byproducts released during manufacture) are greenhouse gases (GHGs), the emissions of which will result in a contribution to the direct positive forcing of the climate. Direct emissions of GHGs may occur during the manufacture of these substances, during their use in products and processes and at the end of the life of these products. Banks are the total amount of substances contained in existing equipment, chemical stockpiles, foams and other products, which are not yet released to the atmosphere. The indirect GHG emissions of applications of ODSs and their replacements are the GHG emissions related to the energy consumption during the entire life cycle of the application.

The emissions of organic gases and ammonia due to the replacement of ODSs in refrigeration and air conditioning are not expected to have significant large-scale impacts on air quality. The local impact of hydrocarbon and ammonia substitutes can be estimated by comparing the anticipated emission to local emissions from all sources. Small but not negligible impacts could occur near localised emission sources and such increases may be of concern, for instance, in areas that currently fail to meet local standards. Persistent degradation products of HFCs and HCFCs (such as trifluoroacetic acid, TFA) are removed from the atmosphere via deposition and washout processes and TFA is toxic to some aquatic life at concentrations approaching 1 mg L^{-1}.

However, degradation of identified sources cannot account for the observed TFA abundances in the oceans, surface waters and atmosphere, indicating that there are larger natural sources of TFA. Current observations show typical concentrations in the ocean of about 0.2 g L^{-1}, while concentrations as high as 40 gL^{-1} have been observed in the Dead Sea and Nevada lakes, suggesting a linkage to salt chemistry. Calculations based on projected HCFC and HFC emissions suggest that the concentration of TFA in rain-water due to their degradation is expected to be between 0.1 g L^{-1} and 0.5 g L^{-1} in the year 2010.

12

Kyoto Protocol to the United Nations Framework Convention on Climate Change

The Parties to this Protocol,

Being Parties to the United Nations Framework Convention on Climate Change, hereinafter referred to as "the Convention",

In pursuit of the ultimate objective of the Convention as stated in its Article 2,

Recalling the provisions of the Convention,

Being guided by Article 3 of the Convention,

Pursuant to the Berlin Mandate adopted by decision 1/CP.1 of the Conference of the Parties to the Convention at its first session,

Have agreed as follows:

Article 1

For the purposes of this Protocol, the definitions contained in Article 1 of the Convention shall apply. In addition:

1. "Conference of the Parties" means the Conference of the Parties to the Convention.

2. "Convention" means the United Nations Framework Convention on Climate Change, adopted in New York on 9 May 1992.

3. "Intergovernmental Panel on Climate Change" means the Intergovernmental Panel on Climate Change established in 1988 jointly by the World Meteorological Organization and the United Nations Environment Programme.

4. "Montreal Protocol" means the Montreal Protocol on Substances that Deplete the Ozone Layer, adopted in Montreal on 16 September 1987 and as subsequently adjusted and amended.

5. "Parties present and voting" means Parties present and casting an affirmative or negative vote.

6. "Party" means, unless the context otherwise indicates, a Party to this Protocol.

7. "Party included in Annex I" means a Party included in Annex I to the Convention, as may be amended, or a Party which has made a notification under Article 4, paragraph 2(g), of the Convention.

Article 2

1. Each Party included in Annex I, in achieving its quantified emission limitation and reduction commitments under Article 3, in order to promote sustainable development, shall:

   (a) Implement and/or further elaborate policies and measures in accordance with its national circumstances, such as:

      (i) Enhancement of energy efficiency in relevant sectors of the national economy;

(ii) Protection and enhancement of sinks and reservoirs of greenhouse gases not controlled by the Montreal Protocol, taking into account its commitments under relevant international environmental agreements; promotion of sustainable forest management practices, afforestation and reforestation;

(iii) Promotion of sustainable forms of agriculture in light of climate change considerations;

(iv) Research on, and promotion, development and increased use of, new and renewable forms of energy, of carbon dioxide sequestration technologies and of advanced and innovative environmentally sound technologies;

(v) Progressive reduction or phasing out of market imperfections, fiscal incentives, tax and duty exemptions and subsidies in all greenhouse gas emitting sectors that run counter to the objective of the Convention and application of market instruments;

(vi) Encouragement of appropriate reforms in relevant sectors aimed at promoting policies and measures which limit or reduce emissions of greenhouse gases not controlled by the Montreal Protocol;

(vii) Measures to limit and/or reduce emissions of greenhouse gases not controlled by the Montreal Protocol in the transport sector;

(viii) Limitation and/or reduction of methane emissions through recovery and use in waste management, as well as in the production, transport and distribution of energy;

(b) Cooperate with other such Parties to enhance the individual and combined effectiveness of their policies and measures adopted under this Article, pursuant to Article 4, paragraph 2(e)(i), of the Convention. To this end, these Parties shall take steps to share their experience and exchange information on such policies and measures, including developing ways of improving their comparability, transparency and effectiveness. The Conference of the Parties serving as the meeting of the Parties to this Protocol shall, at its first session or as soon as practicable thereafter, consider ways to facilitate such cooperation, taking into account all relevant information.

2. The Parties included in Annex I shall pursue limitation or reduction of emissions of greenhouse gases not controlled by the Montreal Protocol from aviation and marine bunker fuels, working through the International Civil Aviation Organization and the International Maritime Organization, respectively.

3. The Parties included in Annex I shall strive to implement policies and measures under this Article in such a way as to minimize adverse effects, including the adverse effects of climate change, effects on international trade, and social, environmental and economic impacts on other Parties, especially developing country Parties and in particular those identified in Article 4, paragraphs 8 and 9, of the Convention, taking into account Article 3 of the Convention. The Conference of the Parties serving as the meeting of the Parties to this Protocol may take further action, as appropriate, to promote the implementation of the provisions of this paragraph.

4. The Conference of the Parties serving as the meeting of the Parties to this Protocol, if it decides that it

would be beneficial to coordinate any of the policies and measures in paragraph 1(a) above, taking into account different national circumstances and potential effects, shall consider ways and means to elaborate the coordination of such policies and measures.

ARTICLE 3

1. The Parties included in Annex I shall, individually or jointly, ensure that their aggregate anthropogenic carbon dioxide equivalent emissions of the greenhouse gases listed in Annex A do not exceed their assigned amounts, calculated pursuant to their quantified emission limitation and reduction commitments inscribed in Annex B and in accordance with the provisions of this Article, with a view to reducing their overall emissions of such gases by at least 5 per cent below 1990 levels in the commitment period 2008 to 2012.

2. Each Party included in Annex I shall, by 2005, have made demonstrable progress in achieving its commitments under this Protocol.

3. The net changes in greenhouse gas emissions by sources and removals by sinks resulting from direct human-induced land-use change and forestry activities, limited to afforestation, reforestation and deforestation since 1990, measured as verifiable changes in carbon stocks in each commitment period, shall be used to meet the commitments under this Article of each Party included in Annex I. The greenhouse gas emissions by sources and removals by sinks associated with those activities shall be reported in a transparent and verifiable manner and reviewed in accordance with Articles 7 and 8.

4. Prior to the first session of the Conference of the Parties serving as the meeting of the Parties to this Protocol, each Party included in Annex I shall provide, for consideration by the Subsidiary Body for Scientific and Technological Advice, data to establish its level of carbon stocks in 1990 and to enable an estimate to be made of its changes in carbon stocks in subsequent years. The Conference of the Parties serving as the meeting of the Parties to this Protocol shall, at its first session or as soon as practicable thereafter, decide upon modalities, rules and guidelines as to how, and which, additional human-induced activities related to changes in greenhouse gas emissions by sources and removals by sinks in the agricultural soils and the land-use change and forestry categories shall be added to, or subtracted from, the assigned amounts for Parties included in Annex I, taking into account uncertainties, transparency in reporting, verifiability, the methodological work of the Intergovernmental Panel on Climate Change, the advice provided by the Subsidiary Body for Scientific and Technological Advice in accordance with Article 5 and the decisions of the Conference of the Parties. Such a decision shall apply in the second and subsequent commitment periods. A Party may choose to apply such a decision on these additional human-induced activities for its first commitment period, provided that these activities have taken place since 1990.

5. The Parties included in Annex I undergoing the process of transition to a market economy whose base year or period was established pursuant to decision 9/CP.2 of the Conference of the Parties at its second session shall use that base year or period for the implementation of their commitments under this Article. Any other Party included in Annex I

undergoing the process of transition to a market economy which has not yet submitted its first national communication under Article 12 of the Convention may also notify the Conference of the Parties serving as the meeting of the Parties to this Protocol that it intends to use an historical base year or period other than 1990 for the implementation of its commitments under this Article. The Conference of the Parties serving as the meeting of the Parties to this Protocol shall decide on the acceptance of such notification.

6. Taking into account Article 4, paragraph 6, of the Convention, in the implementation of their commitments under this Protocol other than those under this Article, a certain degree of flexibility shall be allowed by the Conference of the Parties serving as the meeting of the Parties to this Protocol to the Parties included in Annex I undergoing the process of transition to a market economy.

7. In the first quantified emission limitation and reduction commitment period, from 2008 to 2012, the assigned amount for each Party included in Annex I shall be equal to the percentage inscribed for it in Annex B of its aggregate anthropogenic carbon dioxide equivalent emissions of the greenhouse gases listed in Annex A in 1990, or the base year or period determined in accordance with paragraph 5 above, multiplied by five. Those Parties included in Annex I for whom land-use change and forestry constituted a net source of greenhouse gas emissions in 1990 shall include in their 1990 emissions base year or period the aggregate anthropogenic carbon dioxide equivalent emissions by sources minus removals by sinks in 1990 from land-use change for the purposes of calculating their assigned amount.

8. Any Party included in Annex I may use 1995 as its base year for hydrofluorocarbons, perfluorocarbons and sulphur hexafluoride, for the purposes of the calculation referred to in paragraph 7 above.

9. Commitments for subsequent periods for Parties included in Annex I shall be established in amendments to Annex B to this Protocol, which shall be adopted in accordance with the provisions of Article 21, paragraph 7. The Conference of the Parties serving as the meeting of the Parties to this Protocol shall initiate the consideration of such commitments at least seven years before the end of the first commitment period referred to in paragraph 1 above.

10. Any emission reduction units, or any part of an assigned amount, which a Party acquires from another Party in accordance with the provisions of Article 6 or of Article 17 shall be added to the assigned amount for the acquiring Party.

11. Any emission reduction units, or any part of an assigned amount, which a Party transfers to another Party in accordance with the provisions of Article 6 or of Article 17 shall be subtracted from the assigned amount for the transferring Party.

12. Any certified emission reductions which a Party acquires from another Party in accordance with the provisions of Article 12 shall be added to the assigned amount for the acquiring Party.

13. If the emissions of a Party included in Annex I in a commitment period are less than its assigned amount under this Article, this difference shall, on request of that Party, be added to the assigned amount for that Party for subsequent commitment periods.

14. Each Party included in Annex I shall strive to implement the commitments mentioned in paragraph

1 above in such a way as to minimize adverse social, environmental and economic impacts on developing country Parties, particularly those identified in Article 4, paragraphs 8 and 9, of the Convention. In line with relevant decisions of the Conference of the Parties on the implementation of those paragraphs, the Conference of the Parties serving as the meeting of the Parties to this Protocol shall, at its first session, consider what actions are necessary to minimize the adverse effects of climate change and/or the impacts of response measures on Parties referred to in those paragraphs. Among the issues to be considered shall be the establishment of funding, insurance and transfer of technology.

Article 4

1. Any Parties included in Annex I that have reached an agreement to fulfil their commitments under Article 3 jointly, shall be deemed to have met those commitments provided that their total combined aggregate anthropogenic carbon dioxide equivalent emissions of the greenhouse gases listed in Annex A do not exceed their assigned amounts calculated pursuant to their quantified emission limitation and reduction commitments inscribed in Annex B and in accordance with the provisions of Article 3. The respective emission level allocated to each of the Parties to the agreement shall be set out in that agreement.

2. The Parties to any such agreement shall notify the secretariat of the terms of the agreement on the date of deposit of their instruments of ratification, acceptance or approval of this Protocol, or accession thereto. The secretariat shall in turn inform the Parties

and signatories to the Convention of the terms of the agreement.

3. Any such agreement shall remain in operation for the duration of the commitment period specified in Article 3, paragraph 7.

4. If Parties acting jointly do so in the framework of, and together with, a regional economic integration organization, any alteration in the composition of the organization after adoption of this Protocol shall not affect existing commitments under this Protocol. Any alteration in the composition of the organization shall only apply for the purposes of those commitments under Article 3 that are adopted subsequent to that alteration.

5. In the event of failure by the Parties to such an agreement to achieve their total combined level of emission reductions, each Party to that agreement shall be responsible for its own level of emissions set out in the agreement.

6. If Parties acting jointly do so in the framework of, and together with, a regional economic integration organization which is itself a Party to this Protocol, each member State of that regional economic integration organization individually, and together with the regional economic integration organization acting in accordance with Article 24, shall, in the event of failure to achieve the total combined level of emission reductions, be responsible for its level of emissions as notified in accordance with this Article.

Article 5

1. Each Party included in Annex I shall have in place, no later than one year prior to the start of the first

commitment period, a national system for the estimation of anthropogenic emissions by sources and removals by sinks of all greenhouse gases not controlled by the Montreal Protocol. Guidelines for such national systems, which shall incorporate the methodologies specified in paragraph 2 below, shall be decided upon by the Conference of the Parties serving as the meeting of the Parties to this Protocol at its first session.

2. Methodologies for estimating anthropogenic emissions by sources and removals by sinks of all greenhouse gases not controlled by the Montreal Protocol shall be those accepted by the Intergovernmental Panel on Climate Change and agreed upon by the Conference of the Parties at its third session. Where such methodologies are not used, appropriate adjustments shall be applied according to methodologies agreed upon by the Conference of the Parties serving as the meeting of the Parties to this Protocol at its first session. Based on the work of, inter alia, the Intergovernmental Panel on Climate Change and advice provided by the Subsidiary Body for Scientific and Technological Advice, the Conference of the Parties serving as the meeting of the Parties to this Protocol shall regularly review and, as appropriate, revise such methodologies and adjustments, taking fully into account any relevant decisions by the Conference of the Parties. Any revision to methodologies or adjustments shall be used only for the purposes of ascertaining compliance with commitments under Article 3 in respect of any commitment period adopted subsequent to that revision.

3. The global warming potentials used to calculate the carbon dioxide equivalence of anthropogenic

emissions by sources and removals by sinks of greenhouse gases listed in Annex A shall be those accepted by the Intergovernmental Panel on Climate Change and agreed upon by the Conference of the Parties at its third session. Based on the work of, inter alia, the Intergovernmental Panel on Climate Change and advice provided by the Subsidiary Body for Scientific and Technological Advice, the Conference of the Parties serving as the meeting of the Parties to this Protocol shall regularly review and, as appropriate, revise the global warming potential of each such greenhouse gas, taking fully into account any relevant decisions by the Conference of the Parties. Any revision to a global warming potential shall apply only to commitments under Article 3 in respect of any commitment period adopted subsequent to that revision.

Article 6

1. For the purpose of meeting its commitments under Article 3, any Party included in Annex I may transfer to, or acquire from, any other such Party emission reduction units resulting from projects aimed at reducing anthropogenic emissions by sources or enhancing anthropogenic removals by sinks of greenhouse gases in any sector of the economy, provided that:

   (a) Any such project has the approval of the Parties involved;

   (b) Any such project provides a reduction in emissions by sources, or an enhancement of removals by sinks, that is additional to any that would otherwise occur;

(c) It does not acquire any emission reduction units if it is not in compliance with its obligations under Articles 5 and 7; and

(d) The acquisition of emission reduction units shall be supplemental to domestic actions for the purposes of meeting commitments under Article 3.

2. The Conference of the Parties serving as the meeting of the Parties to this Protocol may, at its first session or as soon as practicable thereafter, further elaborate guidelines for the implementation of this Article, including for verification and reporting.

3. A Party included in Annex I may authorize legal entities to participate, under its responsibility, in actions leading to the generation, transfer or acquisition under this Article of emission reduction units.

4. If a question of implementation by a Party included in Annex I of the requirements referred to in this Article is identified in accordance with the relevant provisions of Article 8, transfers and acquisitions of emission reduction units may continue to be made after the question has been identified, provided that any such units may not be used by a Party to meet its commitments under Article 3 until any issue of compliance is resolved.

Article 7

1. Each Party included in Annex I shall incorporate in its annual inventory of anthropogenic emissions by sources and removals by sinks of greenhouse gases not controlled by the Montreal Protocol, submitted in accordance with the relevant decisions of the

Conference of the Parties, the necessary supplementary information for the purposes of ensuring compliance with Article 3, to be determined in accordance with paragraph 4 below.

2. Each Party included in Annex I shall incorporate in its national communication, submitted under Article 12 of the Convention, the supplementary information necessary to demonstrate compliance with its commitments under this Protocol, to be determined in accordance with paragraph 4 below.

3. Each Party included in Annex I shall submit the information required under paragraph 1 above annually, beginning with the first inventory due under the Convention for the first year of the commitment period after this Protocol has entered into force for that Party. Each such Party shall submit the information required under paragraph 2 above as part of the first national communication due under the Convention after this Protocol has entered into force for it and after the adoption of guidelines as provided for in paragraph 4 below. The frequency of subsequent submission of information required under this Article shall be determined by the Conference of the Parties serving as the meeting of the Parties to this Protocol, taking into account any timetable for the submission of national communications decided upon by the Conference of the Parties.

4. The Conference of the Parties serving as the meeting of the Parties to this Protocol shall adopt at its first session, and review periodically thereafter, guidelines for the preparation of the information required under this Article, taking into account guidelines for the preparation of national communications by Parties included in Annex I adopted by the Conference of the Parties. The Conference of the Parties serving as the

meeting of the Parties to this Protocol shall also, prior to the first commitment period, decide upon modalities for the accounting of assigned amounts.

Article 8

1. The information submitted under Article 7 by each Party included in Annex I shall be reviewed by expert review teams pursuant to the relevant decisions of the Conference of the Parties and in accordance with guidelines adopted for this purpose by the Conference of the Parties serving as the meeting of the Parties to this Protocol under paragraph 4 below. The information submitted under Article 7, paragraph 1, by each Party included in Annex I shall be reviewed as part of the annual compilation and accounting of emissions inventories and assigned amounts. Additionally, the information submitted under Article 7, paragraph 2, by each Party included in Annex I shall be reviewed as part of the review of communications.

2. Expert review teams shall be coordinated by the secretariat and shall be composed of experts selected from those nominated by Parties to the Convention and, as appropriate, by intergovernmental organizations, in accordance with guidance provided for this purpose by the Conference of the Parties.

3. The review process shall provide a thorough and comprehensive technical assessment of all aspects of the implementation by a Party of this Protocol. The expert review teams shall prepare a report to the Conference of the Parties serving as the meeting of the Parties to this Protocol, assessing the implementation of the commitments of the Party and identifying any potential problems in, and factors

influencing, the fulfilment of commitments. Such reports shall be circulated by the secretariat to all Parties to the Convention. The secretariat shall list those questions of implementation indicated in such reports for further consideration by the Conference of the Parties serving as the meeting of the Parties to this Protocol.

4. The Conference of the Parties serving as the meeting of the Parties to this Protocol shall adopt at its first session, and review periodically thereafter, guidelines for the review of implementation of this Protocol by expert review teams taking into account the relevant decisions of the Conference of the Parties.

5. The Conference of the Parties serving as the meeting of the Parties to this Protocol shall, with the assistance of the Subsidiary Body for Implementation and, as appropriate, the Subsidiary Body for Scientific and Technological Advice, consider:

   (a) The information submitted by Parties under Article 7 and the reports of the expert reviews there on conducted under this Article; and

   (b) Those questions of implementation listed by the secretariat under paragraph 3 above, as well as any questions raised by Parties.

6. Pursuant to its consideration of the information referred to in paragraph 5 above, the Conference of the Parties serving as the meeting of the Parties to this Protocol shall take decisions on any matter required for the implementation of this Protocol.

Article 9

1. The Conference of the Parties serving as the meeting of the Parties to this Protocol shall periodically review

this Protocol in the light of the best available scientific information and assessments on climate change and its impacts, as well as relevant technical, social and economic information. Such reviews shall be coordinated with pertinent reviews under the Convention, in particular those required by Article 4, paragraph 2(d), and Article 7, paragraph 2(a), of the Convention. Based on these reviews, the Conference of the Parties serving as the meeting of the Parties to this Protocol shall take appropriate action.

2. The first review shall take place at the second session of the Conference of the Parties serving as the meeting of the Parties to this Protocol. Further reviews shall take place at regular intervals and in a timely manner.

Article 10

All Parties, taking into account their common but differentiated responsibilities and their specific national and regional development priorities, objectives and circumstances, without introducing any new commitments for Parties not included in Annex I, but reaffirming existing commitments under Article 4, paragraph 1, of the Convention, and continuing to advance the implementation of these commitments in order to achieve sustainable development, taking into account Article 4, paragraphs 3, 5 and 7, of the Convention, shall:

(a) Formulate, where relevant and to the extent possible, cost-effective national and, where appropriate, regional programmes to improve the quality of local emission factors, activity data and/or models which reflect the socio-economic conditions of each Party for the preparation and periodic updating of national inventories of anthropogenic emissions by sources and removals by sinks of all greenhouse gases not controlled by the Montreal Protocol, using comparable methodologies to be agreed upon by the Conference of the Parties, and consistent with the guidelines for the preparation of national communications adopted by the Conference of the Parties;

(b) Formulate, implement, publish and regularly update national and, where appropriate, regional programmes containing measures to mitigate climate change and measures to facilitate adequate adaptation to climate change:

(i) Such programmes would, inter alia, concern the energy, transport and industry sectors as well as agriculture, forestry and waste management. Furthermore, adaptation technologies and methods for improving spatial planning would improve adaptation to climate change; and

(ii) Parties included in Annex I shall submit information on action under this Protocol, including national programmes, in accordance with Article 7; and other Parties shall seek to include in their national communications, as appropriate, information on programmes which contain measures that the Party believes contribute to addressing climate change and its adverse impacts, including the abatement of increases in greenhouse gas emissions, and enhancement of and removals by sinks, capacity building and adaptation measures;

(c) Cooperate in the promotion of effective modalities for the development, application and diffusion of, and take all practicable steps to promote, facilitate and finance, as appropriate, the transfer of, or access to, environmentally sound technologies, know-how, practices and processes pertinent to climate change, in particular to developing countries, including the formulation of policies and programmes for the effective transfer of environmentally sound technologies that are publicly owned or in the public domain and the creation of an enabling environment for the private sector, to promote and enhance the transfer of, and access to, environmentally sound technologies;

(d) Cooperate in scientific and technical research and promote the maintenance and the development of systematic observation systems and development of data archives to reduce uncertainties related to the climate system, the adverse impacts of climate change and the economic and social consequences of various response strategies, and promote the development and strengthening of endogenous capacities and capabilities to participate in international and intergovernmental efforts, programmes and networks on research and systematic observation, taking into account Article 5 of the Convention;

(e) Cooperate in and promote at the international level, and, where appropriate, using existing bodies, the development and implementation of education and training programmes, including the strengthening of national capacity building, in particular human and institutional capacities and the exchange or secondment of personnel to train experts in this field, in particular for developing countries, and facilitate at the national level public awareness of, and public access to information on, climate change. Suitable modalities should be developed to implement these activities through the relevant bodies of the Convention, taking into account Article 6 of the Convention;

(f) Include in their national communications information on programmes and activities undertaken pursuant to this Article in accordance with relevant decisions of the Conference of the Parties; and

(g) Give full consideration, in implementing the commitments under this Article, to Article 4, paragraph 8, of the Convention.

Article 11

1. In the implementation of Article 10, Parties shall take into account the provisions of Article 4, paragraphs 4, 5, 7, 8 and 9, of the Convention.

2. In the context of the implementation of Article 4, paragraph 1, of the Convention, in accordance with the provisions of Article 4, paragraph 3, and Article 11 of the Convention, and through the entity or entities entrusted with the operation of the financial mechanism of the Convention, the developed country Parties and other developed Parties included in Annex II to the Convention shall:

   (a) Provide new and additional financial resources to meet the agreed full costs incurred by developing country Parties in advancing the implementation of existing commitments under Article 4, paragraph 1(a), of the Convention that are covered in Article 10, subparagraph (a); and

   (b) Also provide such financial resources, including for the transfer of technology, needed by the developing country Parties to meet the agreed full incremental costs of advancing the implementation of existing commitments under Article 4, paragraph 1, of the Convention that are covered by Article 10 and that are agreed between a developing country Party and the international entity or entities referred to in Article 11 of the Convention, in accordance with that Article.

The implementation of these existing commitments shall take into account the need for adequacy and predictability in the flow of funds and the importance of appropriate burden sharing among developed country Parties. The guidance to the entity or entities entrusted with the operation of the financial mechanism of the Convention in relevant decisions of the Conference of the Parties, including those agreed before the adoption of this Protocol, shall apply mutatis mutandis to the provisions of this paragraph.

3. The developed country Parties and other developed Parties in Annex II to the Convention may also provide, and developing country Parties avail themselves of, financial resources for the implementation of Article 10, through bilateral, regional and other multilateral channels.

Article 12

1. A clean development mechanism is hereby defined.
2. The purpose of the clean development mechanism shall be to assist Parties not included in Annex I in achieving sustainable development and in contributing to the ultimate objective of the Convention, and to assist Parties included in Annex I in achieving compliance with their quantified emission limitation and reduction commitments under Article 3.
3. Under the clean development mechanism:

   (a) Parties not included in Annex I will benefit from project activities resulting in certified emission reductions; and

(b) Parties included in Annex I may use the certified emission reductions accruing from such project activities to contribute to compliance with part of their quantified emission limitation and reduction commitments under Article 3, as determined by the Conference of the Parties serving as the meeting of the Parties to this Protocol.

4. The clean development mechanism shall be subject to the authority and guidance of the Conference of the Parties serving as the meeting of the Parties to this Protocol and be supervised by an executive board of the clean development mechanism.

5. Emission reductions resulting from each project activity shall be certified by operational entities to be designated by the Conference of the Parties serving as the meeting of the Parties to this Protocol, on the basis of:

(a) Voluntary participation approved by each Party involved;

(b) Real, measurable, and long-term benefits related to the mitigation of climate change; and

(c) Reductions in emissions that are additional to any that would occur in the absence of the certified project activity.

6. The clean development mechanism shall assist in arranging funding of certified project activities as necessary.

7. The Conference of the Parties serving as the meeting of the Parties to this Protocol shall, at its first session, elaborate modalities and procedures with the objective of ensuring transparency, efficiency and accountability through independent auditing and verification of project activities.

8. The Conference of the Parties serving as the meeting of the Parties to this Protocol shall ensure that a share of the proceeds from certified project activities is used to cover administrative expenses as well as to assist developing country Parties that are particularly vulnerable to the adverse effects of climate change to meet the costs of adaptation.

9. Participation under the clean development mechanism, including in activities mentioned in paragraph 3(a) above and in the acquisition of certified emission reductions, may involve private and/or public entities, and is to be subject to whatever guidance may be provided by the executive board of the clean development mechanism.

10. Certified emission reductions obtained during the period from the year 2000 up to the beginning of the first commitment period can be used to assist in achieving compliance in the first commitment period.

Article 13

1. The Conference of the Parties, the supreme body of the Convention, shall serve as the meeting of the Parties to this Protocol.

2. Parties to the Convention that are not Parties to this Protocol may participate as observers in the proceedings of any session of the Conference of the Parties serving as the meeting of the Parties to this Protocol. When the Conference of the Parties serves as the meeting of the Parties to this Protocol, decisions under this Protocol shall be taken only by those that are Parties to this Protocol.

3. When the Conference of the Parties serves as the meeting of the Parties to this Protocol, any member

of the Bureau of the Conference of the Parties representing a Party to the Convention but, at that time, not a Party to this Protocol, shall be replaced by an additional member to be elected by and from amongst the Parties to this Protocol.

4. The Conference of the Parties serving as the meeting of the Parties to this Protocol shall keep under regular review the implementation of this Protocol and shall make, within its mandate, the decisions necessary to promote its effective implementation. It shall perform the functions assigned to it by this Protocol and shall:

   (a) Assess, on the basis of all information made available to it in accordance with the provisions of this Protocol, the implementation of this Protocol by the Parties, the overall effects of the measures taken pursuant to this Protocol, in particular environmental, economic and social effects as well as their cumulative impacts and the extent to which progress towards the objective of the Convention is being achieved;

   (b) Periodically examine the obligations of the Parties under this Protocol, giving due consideration to any reviews required by Article 4, paragraph 2(d), and Article 7, paragraph 2, of the Convention, in the light of the objective of the Convention, the experience gained in its implementation and the evolution of scientific and technological knowledge, and in this respect consider and adopt regular reports on the implementation of this Protocol;

   (c) Promote and facilitate the exchange of information on measures adopted by the Parties to address climate change and its effects, taking into account the differing circumstances, responsibilities and capabilities of the Parties and their respective commitments under this Protocol;

(d) Facilitate, at the request of two or more Parties, the coordination of measures adopted by them to address climate change and its effects, taking into account the differing circumstances, responsibilities and capabilities of the Parties and their respective commitments under this Protocol;

(e) Promote and guide, in accordance with the objective of the Convention and the provisions of this Protocol, and taking fully into account the relevant decisions by the Conference of the Parties, the development and periodic refinement of comparable methodologies for the effective implementation of this Protocol, to be agreed on by the Conference of the Parties serving as the meeting of the Parties to this Protocol;

(f) Make recommendations on any matters necessary for the implementation of this Protocol;

(g) Seek to mobilize additional financial resources in accordance with Article 11, paragraph 2;

(h) Establish such subsidiary bodies as are deemed necessary for the implementation of this Protocol;

(i) Seek and utilize, where appropriate, the services and cooperation of, and information provided by, competent international organizations and intergovernmental and non-governmental bodies; and

(j) Exercise such other functions as may be required for the implementation of this Protocol, and consider any assignment resulting from a decision by the Conference of the Parties.

5. The rules of procedure of the Conference of the Parties and financial procedures applied under the Convention shall be applied mutatis mutandis under this Protocol, except as may be otherwise decided by consensus by the Conference of the Parties serving as the meeting of the Parties to this Protocol.

6. The first session of the Conference of the Parties serving as the meeting of the Parties to this Protocol shall be convened by the secretariat in conjunction with the first session of the Conference of the Parties that is scheduled after the date of the entry into force of this Protocol. Subsequent ordinary sessions of the Conference of the Parties serving as the meeting of the Parties to this Protocol shall be held every year and in conjunction with ordinary sessions of the Conference of the Parties, unless otherwise decided by the Conference of the Parties serving as the meeting of the Parties to this Protocol.

7. Extraordinary sessions of the Conference of the Parties serving as the meeting of the Parties to this Protocol shall be held at such other times as may be deemed necessary by the Conference of the Parties serving as the meeting of the Parties to this Protocol, or at the written request of any Party, provided that, within six months of the request being communicated to the Parties by the secretariat, it is supported by at least one third of the Parties.

8. The United Nations, its specialized agencies and the International Atomic Energy

Agency, as well as any State member thereof or observers thereto not party to the Convention, may be represented at sessions of the Conference of the Parties serving as the meeting of the Parties to this Protocol as observers. Any body or agency, whether national or international, governmental or non-governmental, which is qualified in matters covered by this Protocol and which has informed the secretariat of its wish to be represented at a session of the Conference of the Parties serving as the meeting of the Parties to this Protocol as an observer, may be so admitted unless at least one third of the Parties present object. The admission and participation of observers shall be subject to the rules of procedure, as referred to in paragraph 5 above.

Article 14

1. The secretariat established by Article 8 of the Convention shall serve as the secretariat of this Protocol.

2. Article 8, paragraph 2, of the Convention on the functions of the secretariat, and Article 8, paragraph 3, of the Convention on arrangements made for the functioning of the secretariat, shall apply mutatis mutandis to this Protocol. The secretariat shall, in addition, exercise the functions assigned to it under this Protocol.

Article 15

1. The Subsidiary Body for Scientific and Technological Advice and the Subsidiary Body for Implementation established by Articles 9 and 10 of the Convention shall serve as, respectively, the Subsidiary Body for Scientific and Technological Advice and the Subsidiary Body for Implementation of this Protocol. The provisions relating to the functioning of these two bodies under the Convention shall apply mutatis mutandis to this Protocol. Sessions of the meetings of the Subsidiary Body for Scientific and Technological Advice and the Subsidiary Body for Implementation

of this Protocol shall be held in conjunction with the meetings of, respectively, the Subsidiary Body for Scientific and Technological Advice and the Subsidiary Body for Implementation of the Convention.

2. Parties to the Convention that are not Parties to this Protocol may participate as observers in the proceedings of any session of the subsidiary bodies. When the subsidiary bodies serve as the subsidiary bodies of this Protocol, decisions under this Protocol shall be taken only by those that are Parties to this Protocol.

3. When the subsidiary bodies established by Articles 9 and 10 of the Convention exercise their functions with regard to matters concerning this Protocol, any member of the Bureaux of those subsidiary bodies representing a Party to the Convention but, at that time, not a party to this Protocol, shall be replaced by an additional member to be elected by and from amongst the Parties to this Protocol.

Article 16

The Conference of the Parties serving as the meeting of the Parties to this Protocol shall, as soon as practicable, consider the application to this Protocol of, and modify as appropriate, the multilateral consultative process referred to in Article 13 of the Convention, in the light of any relevant decisions that may be taken by the Conference of the Parties. Any multilateral consultative process that may be applied to this Protocol shall operate without prejudice to the procedures and mechanisms established in accordance with Article 18.

Artic'e 17

The Conference of the Parties shall define the relevant

principles, modalities, rules and guidelines, in particular for verification, reporting and accountability for emissions trading. The Parties included in Annex B may participate in emissions trading for the purposes of fulfilling their commitments under Article 3. Any such trading shall be supplemental to domestic actions for the purpose of meeting quantified emission limitation and reduction commitments under that Article.

ARTICLE 18

The Conference of the Parties serving as the meeting of the Parties to this Protocol shall, at its first session, approve appropriate and effective procedures and mechanisms to determine and to address cases of non-compliance with the provisions of this Protocol, including through the development of an indicative list of consequences, taking into account the cause, type, degree and frequency of non-compliance. Any procedures and mechanisms under this Article entailing binding consequences shall be adopted by means of an amendment to this Protocol.

ARTICLE 19

The provisions of Article 14 of the Convention on settlement of disputes shall apply mutatis mutandis to this Protocol.

ARTICLE 20

1. Any Party may propose amendments to this Protocol.
2. Amendments to this Protocol shall be adopted at an ordinary session of the Conference of the Parties serving as the meeting of the Parties to this Protocol. The text of any proposed amendment to this Protocol shall be communicated to the Parties by the secretariat

at least six months before the meeting at which it is proposed for adoption. The secretariat shall also communicate the text of any proposed amendments to the Parties and signatories to the Convention and, for information, to the Depositary.

3. The Parties shall make every effort to reach agreement on any proposed amendment to this Protocol by consensus. If all efforts at consensus have been exhausted, and no agreement reached, the amendment shall as a last resort be adopted by a three-fourths majority vote of the Parties present and voting at the meeting. The adopted amendment shall be communicated by the secretariat to the Depositary, who shall circulate it to all Parties for their acceptance.

4. Instruments of acceptance in respect of an amendment shall be deposited with the Depositary. An amendment adopted in accordance with paragraph 3 above shall enter into force for those Parties having accepted it on the ninetieth day after the date of receipt by the Depositary of an instrument of acceptance by at least three fourths of the Parties to this Protocol.

5. The amendment shall enter into force for any other Party on the ninetieth day after the date on which that Party deposits with the Depositary its instrument of acceptance of the said amendment.

Article 21

1. Annexes to this Protocol shall form an integral part thereof and, unless otherwise expressly provided, a reference to this Protocol constitutes at the same time a reference to any annexes thereto. Any annexes adopted after the entry into force of this Protocol shall

be restricted to lists, forms and any other material of a descriptive nature that is of a scientific, technical, procedural or administrative character.

2. Any Party may make proposals for an annex to this Protocol and may propose amendments to annexes to this Protocol.

3. Annexes to this Protocol and amendments to annexes to this Protocol shall be adopted at an ordinary session of the Conference of the Parties serving as the meeting of the Parties to this Protocol. The text of any proposed annex or amendment to an annex shall be communicated to the Parties by the secretariat at least six months before the meeting at which it is proposed for adoption. The secretariat shall also communicate the text of any proposed annex or amendment to an annex to the Parties and signatories to the Convention and, for information, to the Depositary.

4. The Parties shall make every effort to reach agreement on any proposed annex or amendment to an annex by consensus. If all efforts at consensus have been exhausted, and no agreement reached, the annex or amendment to an annex shall as a last resort be adopted by a three-fourths majority vote of the Parties present and voting at the meeting. The adopted annex or amendment to an annex shall be communicated by the secretariat to the Depositary, who shall circulate it to all Parties for their acceptance.

5. An annex, or amendment to an annex other than Annex A or B, that has been adopted in accordance with paragraphs 3 and 4 above shall enter into force for all Parties to this Protocol six months after the date of the communication by the Depositary to such Parties of the adoption of the annex or adoption of the amendment to the annex, except for those Parties

that have notified the Depositary, in writing, within that period of their non-acceptance of the annex or amendment to the annex. The annex or amendment to an annex shall enter into force for Parties which withdraw their notification of non-acceptance on the ninetieth day after the date on which withdrawal of such notification has been received by the Depositary.

6. If the adoption of an annex or an amendment to an annex involves an amendment to this Protocol, that annex or amendment to an annex shall not enter into force until such time as the amendment to this Protocol enters into force.

7. Amendments to Annexes A and B to this Protocol shall be adopted and enter into force in accordance with the procedure set out in Article 20, provided that any amendment to Annex B shall be adopted only with the written consent of the Party concerned.

Article 22

1. Each Party shall have one vote, except as provided for in paragraph 2 below.

2. Regional economic integration organizations, in matters within their competence, shall exercise their right to vote with a number of votes equal to the number of their member States that are Parties to this Protocol. Such an organization shall not exercise its right to vote if any of its member States exercises its right, and vice versa.

Article 23

The Secretary-General of the United Nations shall be the Depositary of this Protocol.

Article 24

1. This Protocol shall be open for signature and subject to ratification, acceptance or approval by States and regional economic integration organizations which are Parties to the Convention. It shall be open for signature at United Nations Headquarters in New York from 16 March 1998 to 15 March 1999. This Protocol shall be open for accession from the day after the date on which it is closed for signature. Instruments of ratification, acceptance, approval or accession shall be deposited with the Depositary.
2. Any regional economic integration organization which becomes a Party to this Protocol without any of its member States being a Party shall be bound by all the obligations under this Protocol. In the case of such organizations, one or more of whose member States is a Party to this Protocol, the organization and its member States shall decide on their respective responsibilities for the performance of their obligations under this Protocol. In such cases, the organization and the member States shall not be entitled to exercise rights under this Protocol concurrently.
3. In their instruments of ratification, acceptance, approval or accession, regional economic integration organizations shall declare the extent of their competence with respect to the matters governed by this Protocol. These organizations shall also inform the Depositary, who shall in turn inform the Parties, of any substantial modification in the extent of their competence.

Article 25

1. This Protocol shall enter into force on the ninetieth

day after the date on which not less than 55 Parties to the Convention, incorporating Parties included in Annex I which accounted in total for at least 55 per cent of the total carbon dioxide emissions for 1990 of the Parties included in Annex I, have deposited their instruments of ratification, acceptance, approval or accession.

2. For the purposes of this Article, "the total carbon dioxide emissions for 1990 of the Parties included in Annex I" means the amount communicated on or before the date of adoption of this Protocol by the Parties included in Annex I in their first national communications submitted in accordance with Article 12 of the Convention.

3. For each State or regional economic integration organization that ratifies, accepts or approves this Protocol or accedes thereto after the conditions set out in paragraph 1 above for entry into force have been fulfilled, this Protocol shall enter into force on the ninetieth day following the date of deposit of its instrument of ratification, acceptance, approval or accession.

4. For the purposes of this Article, any instrument deposited by a regional economic integration organization shall not be counted as additional to those deposited by States members of the organization.

Article 26

No reservations may be made to this Protocol.

Article 27

1. At any time after three years from the date on which this Protocol has entered into force for a Party, that

Party may withdraw from this Protocol by giving written notification to the Depositary.

2. Any such withdrawal shall take effect upon expiry of one year from the date of receipt by the Depositary of the notification of withdrawal, or on such later date as may be specified in the notification of withdrawal.

3. Any Party that withdraws from the Convention shall be considered as also having withdrawn from this Protocol.

Article 28

The original of this Protocol, of which the Arabic, Chinese, English, French, Russian and Spanish texts are equally authentic, shall be deposited with the Secretary-General of the United Nations.

DONE at Kyoto this eleventh day of December one thousand nine hundred and ninety-seven.

IN WITNESS WHEREOF the undersigned, being duly authorized to that effect, have affixed their signatures to this Protocol on the dates indicated.

Annex A -

Greenhouse Gases

— Carbon dioxide (CO_2)

— Methane (CH_4)

— Nitrous oxide (N_2O)

— Hydrofluorocarbons (HFCs)

— Perfluorocarbons (PFCs)

— Sulphur hexafluoride (SF_6)

Sectors/Source Categories

— Energy

— Fuel combustion

— Energy industries

— Manufacturing industries and construction

— Transport

— Other sectors

— Other

— Fugitive emissions from fuels

— Solid fuels

— Oil and natural gas

— Other

— Industrial processes

— Mineral products

— Chemical industry

— Metal production

— Other production

— Production of halocarbons and sulphur hexafluoride

— Consumption of halocarbons and sulphur hexafluoride

— Other

— Solvent and other product use

— Agriculture

— Enteric fermentation

— Manure management

— Rice cultivation

— Agricultural soils

— Prescribed burning of savannas

— Field burning of agricultural residues

— Other

— Waste

— Solid waste disposal on land

— Wastewater handling

— Waste incineration

— Other

ANNEX B

Party Quantified emission limitation or reduction commitment

(percentage of base year or period)

— Australia 108

— Austria 92

— Belgium 92

— Bulgaria* 92

— Canada 94

— Croatia* 95

— Czech Republic* 92

— Denmark 92

— Estonia* 92

— European Community 92

— Finland 92

— France 92

— Germany 92

— Greece 92

— Hungary* 94

— Iceland 110

— Ireland 92

— Italy 92

— Japan 94

— Latvia* 92

— Liechtenstein 92

— Lithuania* 92

— Luxembourg 92

— Monaco 92

— Netherlands 92

— New Zealand 100

— Norway 101

— Poland* 94

— Portugal 92

— Romania* 92

— Russian Federation* 100

— Slovakia* 92

— Slovenia* 92

— Spain 92

— Sweden 92

— Switzerland 92

— Ukraine* 100

— United Kingdom of Great Britain and Northern Ireland 92

— United States of America 93

* Countries that are undergoing the process of transition to a market economy.

Bibliography

Abrahamson, Dean Edwin, ed. *The Challenge of Global Warming*. Washington, D.C.: Island Press, c1989.

Alverson, Keith D. and Thomas Pedersen. *Environmental Variability and Climate Change*. [Stockholm]: International Geosphere-Biosphere Programme, 2001.

Edgerton, Lynne T. *The Rising Tide: Global Warming and World Sea Levels*. Washington, D.C.: Island Press, c1991.

Gates, David Murray. *Climate Change and its Biological Consequences*. Sunderland, MA: Sinauer Associates, c1993.

International Geosphere-Biosphere Programme. *A Study of Global Change*. Stockholm, Sweden: IGBP Secretariat, Royal Swedish Academy of Sciences, 1998.

Minger, Terrell J. *Greenhouse Glasnost: The Crisis of Global Warming: Essays*. New York: Ecco Press; [Salt Lake City, Utah]: Institute for Resource Management, c1990.

Nunn, Patrick D. *Environmental Change in the Pacific Basin: Chronologies, Causes, Consequences*. Chichester, West Sussex, England; New York: Wiley, c1999.

Peters, Robert L. and Thomas E. Lovejoy, eds. *Global Warming and Biological Diversity*. New Haven, CT: Yale University Press, c1992..

Somerville, Richard. *The Forgiving Air: Understanding Environmental Change*. Berkeley, CA: University of California Press, c1996. Biodiv QC879.7 .S66 1996.

Steffan, Will and Peter Tyson, eds. *Global Change and the Earth System: A Planet Under Pressure*. [Stockholm]: International Geosphere-Biosphere Programme, c2001.

Thompson, Russell D. and Allen Perry, eds. *Applied Climatology: Principles and Practice*. London; New York: Routledge, 1997.

Veroustraete, F., et al., eds. *Vegetation, Modelling and Climatic Change Effects*. The Hague, The Netherlands: SPB Academic Pub., c1994.

Wright, H. E., Jr., et al., eds. *Global Climates Since the Last Glacial Maximum*. Minneapolis, MN: University of Minnesota Press, c1993.